LIVE YOUR PASSION

LIVE YOUR PASSION

BUILDING A WATCH MANUFACTURER

FREDERIQUE CONSTANT . ALPINA . DEMONACO

ALETTA & PETER STAS . GISBERT BRUNNER & ALEXANDER LINZ

Schiffer Publishing Ltd

4880 Lower Valley Road • Atglen, PA 19310

Originally published as:
Lebe Deine Leidenschaft | Live Your Passion
by Ebner Verlag GmbH & Co KG, Ulm
Karlstr. 41, 89073 Ulm
Germany

Library of Congress Control Number: 2014931213

Type set in Aire Bold Pro and Jigsaw Light

ISBN: 978-0-7643-4616-3
Printed in China

Published by Schiffer Publishing, Ltd.
4880 Lower Valley Road
Atglen, PA 19310
Phone: (610) 593-1777; Fax: (610) 593-2002
E-mail: Info@schifferbooks.com

CONTENTS

FREDERIQUE CONSTANT

ALPINA

ATELIERS DEMONACO

LIVE YOUR PASSION

BUILDING A WATCH MANUFACTURER

It all started with our passion for watches. Then, the passion of our watchmakers was added, and since then, we have been on a journey to build a thriving watch manufacture. With our slogan "Live your passion," we encourage others to share in our passion for fine watchmaking.

True passion has always been the result of fine sensitive processes that evolve over time - not the exploits of some exciting moment. Passion represents our internal drive, not our daily habits or needs. While our habits change, our drive continues to be the result of our character and heritage. Time and moments pass. The accomplishments of our drive remain over time. This has been the true aspect of our passion at all times.

ALETTA & PETER STAS
COO & CEO of Frédérique Constant

The idea for this book was born during a seminar on the "Rockefeller Habits" in Amsterdam, which we attended with our company's management team. In the past, various people - journalists, customers, suppliers and industry colleagues - had suggested writing a book about both our passion and the history of Frédérique Constant. It was the suggestions from our Management Team that made us think more seriously about the idea.

Initially, the primary purpose was to create a book as a publicity and information vehicle for potential customers and new employees. We wanted to explain how and why the company came into this world. We certainly also have an ambition to leave some kind of a legacy, to leave behind a "rule book" on what we envision as the long-term culture of the company. We would like this book to inspire and motivate those working with us to continue the growth of The Frédérique Constant, Alpina and Ateliers deMonaco brands. Last but not least, our intention in writing this book is also to give our children a more in-depth understanding of what their parents have been spending so much time on over the past twenty years.

We were told at the YPO workshop, "Write That Book," that it takes one thousand hours to develop and finish a book manuscript. Considering the time we were already spending at work, this hardly sounded like an appealing proposition. The book idea was officially postponed to a later date when we would have more time.

Still, the desire to write something more comprehensive remained. Writing press articles and blog entries did relieve some pressure, but it was never enough. So, quietly, we began to develop the outline of a book. With the help of Gisbert Brunner and Alexander Linz, we were able to form the semblance of a structure out of scattered thoughts and information.

This publication is a mixture between a "Coffee Table Book" and a "Business Book." You will find ample text on the business aspects of the company, our goal to "Let more people enjoy luxury" and the company's Core Values. You also will find a wide variety of photos and images that tell the story of the creation of watch calibers, models, and their manufacturing. Most important, however, is the story of the people in our company: its customers and suppliers. We are very grateful to have been able to work with many excellent people over the past twenty years. Without them, the company would never have been possible.

The following pages begin with an in-depth explanation of the concept of "Accessible Luxury" that was the basis of the creation of Frédérique Constant more than twenty years ago. We explain many details and share specific anecdotes over the following few chapters in interview format. Individual company Core Values are described later, including interviews with colleagues regarding each respective core value.

We have often been asked the reason for the success of our company:

Tell me, what is the single most important success factor?

Unfortunately, or fortunately, there is no single answer to that complex question. A combination of many factors created success at the various development phases of the company. While it is very important to have a clear strategy, we are truly of the opinion that you must do many things right to be successful. That means hard work and long hours. You can read all about it in the rest of this book!

Frédérique Constant watches are, first and foremost, very classic. Over the years, various Frédérique Constant customers asked us to develop a sports collection. However, we always viewed this notion as conflicting with our initial strategy to create, produce, and market classic watches. For us, it was essential to stay consistent with the original, classic strategy. In 2002, we had the opportunity to acquire Alpina, a watch company founded in 1883, with a long history in sports watches. This acquisition gave us a vehicle for sports and lifestyle watches.

Later chapters describe the extensive history of the "Union Horlogere," the company that created and registered Alpina in 1901, and how the company revised the "Alpinist" concept of collaboration. Today, we remain on a mission to revive Alpina to its former glory as a leading world watch brand. It is a work in progress.

Ateliers deMonaco was incorporated in April, 2009, and is the third watch company in our group. Ateliers deMonaco creates and produces high-end watch complications, including a Minute-Repeater, Perpetual Calendar and various Tourbillons. It is in this company that our most talented watchmakers can express themselves and gain experience, while marketing their unique watch creations. Ateliers deMonaco is ultimately an implementation of our "Personal Development" Core Value.

Pim Koeslag and Manuel da Silva will explain how and why they created various, extremely complicated watch mechanisms, as well as how they obtained patents for them and the long process of prototyping and production. Most importantly, they will explain their passion and vision for the future.

We would like to thank everyone who has been involved with Frédérique Constant, Alpina and Ateliers deMonaco, as well as those who have participated in the creation of this book. We thank all those who agreed to be interviewed: Marie Anthonioz, Pim Koeslag, Alexandra Sanchez, Benoit Favrat, José Dos Santos, Hugo Herrero-Levy, Manuel da Silva and Olivier Zaugg. We thank Gisbert Brunner and Alexander Linz for their invaluable contributions to this book.

We also especially thank you for your interest in our watch manufacturer, and we wish you great pleasure while reading about our passion.

Aletta & Peter Stas
Geneva, October 2012

IT ALL BEGAN WITH THE IDEA

FRÉDÉRIQUE CONSTANT

Anyone involved in the manufacturing of watches must, by necessity, have an affinity for this complex subject. How did you find your way to watches, Aletta?
ALETTA: I first encountered watches when I was five years old. My father wasn't in the watch business, but he travelled around the world and often received gifts. That's how I got my first wristwatch, a Timex. It was a hand-wound watch, but I wasn't yet aware of such distinctions. Later, I was given other watches that my father brought home from his travels. For example, he gave me a watch from Japan with a hologram on its dial.

Were those mostly Japanese watches?
ALETTA: Timex was 100% USA at the time.

What did you find particularly fascinating about the watches?
ALETTA: I was less interested in the inner life and more interested in the design. I clearly recall those early encounters with watches because I was asked about them later by classmates who had known me as a child and teen. They hadn't been especially interested in watches, but they were aware of my fascination. Now that we're in business with Frédérique Constant, old friends and acquaintances often raise the subject of bygone days.

What did you wear on your wrist before you started Frédérique Constant, Peter?
PETER: A Seiko diver's watch and a very slim Baume & Mercier watch with a silver-colored dial. Then we bought a Bucherer watch and it triggered our entrepreneurial ambitions.

Please tell me about that, Peter.
PETER: We were on a skiing holiday in Crans Montana, where Aletta's parents have an appartement. In Bern, passing Bucherer's display window, I saw a very beautiful wristwatch with the jeweler's insignia on its dial. Bucherer still made relatively simple watches in those days. They were essentially accessible versions of watches from renowned brands.

And you were there too, Aletta?
ALETTA: Of course. We traveled to Bern together and bought that watch for Peter. It was very classical and if it hadn't borne that insignia, it could have been mistaken for something much more costly.

Quartz or mechanical?
ALETTA: A self-winding mechanical watch with moonphase display. It cost only about one-fifth as much as other brands charged for similar products. If I recall correctly, it also had a transparent back. It was a very attractive timepiece and it looked great on Peter's wrist.

What did this purchase trigger for you?
PETER: That wristwatch sparked our initial ideas. We were still young and unmarried, and we didn't have a big budget.

So it was the exactly right watch at the right time?
PETER: Absolutely. It always makes us happy to think about it..

Were you already familiar with all of the important watch brands back then?
ALETTA: By no means. We knew some brands, but certainly not all. Watches weren't a very big topic in Holland in those days.

Ski vacation in Crans Montana

Did Peter regularly wear his new watch?
PETER: Yes, because I was enthusiastic about it...

ALETTA: ... and because we kept asking ourselves how a manufacturer could possibly sell this product at such an appealingly low price. Of course, we had seen other watches with similar appearances, but much costlier price tags. We knew that Bucherer is a real luxury brand which would never put its name on mediocre merchandise.

PETER: But we also knew that Bucherer's name is associated with watch dealerships and that Bucherer would therefore find it difficult to develop into a global brand that would also be distributed by other jewelers.

Bucherer isn't the only example.
ALETTA: That's right. Just think of Wempe. They're probably in a similar situation. Wempe is quite well known in Germany and perhaps also in the USA, where the company operates some shops, but otherwise that form of business structure can hardly work.

PETER: One model alone isn't likely to be very successful, but that ultimately doesn't matter. Much more important was that we had noticed something in Bern which we could expend into a business model.

Is that when the idea of watches became firmly planted in your minds?
ALETTA: Yes. Peter's wristwatch was a constant reminder. And we really liked the idea. That didn't change when we moved to Hong Kong.

PETER: Quite the opposite: watches were much more visibly present there than in Europe. It seemed as though streets throughout the city were lined door-to-door with watch shops.

What about your past together?
Where did you first meet?
ALETTA: At a party for students in Brussels.

Would "love at first sight" be an appropriate phrase?
ALETTA: Yes, you could call it that.

PETER: I agree wholeheartedly!

ALETTA: I had studied law in Leiden and earned a master's degree after completing a rather economically oriented curriculum in commercial law. I wasn't very interested in ordinary jurisprudence.

PETER: Economic aspects were also in the foreground for me. That's why I studied business management at Erasmus University in Rotterdam.

ALETTA: When we got to know one another better, it soon became clear that Peter has a genuinely entrepreneurial mindset.

How did you notice that?
ALETTA: Peter had just received his pilot's license and he wanted to do something with it, so we considered the idea of shooting aerial photos of houses and selling the pictures for a profit.

PETER: Aletta had been involved in real-estate in the course of her career, so it was only natural for us to think about a project of this kind.

ALETTA: We developed brochures at ING for commercial real-estate projects. To include aerial photos in those brochures was regarded as a modern and appealing touch. That's exactly what we did. (1)

PETER: Once we flew from Rotterdam to Utrecht, found the house, photographed it...

ALETTA: ... and my company was so impressed by our photo that they bought it.

Not every reader is familiar with ING.
ALETTA: When I began there, ING was a split business with two focal points: a bank and an insurance company. In Holland, I began at the insurer, in its department for commercial real-estate. Things were much more dynamic there than in the somewhat listless insurance department. I really enjoyed the dynamism.

Let's get back to flying. What sort of pilot's license had Peter earned?
ALETTA: A license for small single-engine aircraft with a maximum of four seats – Cessnas, for example.

PETER: But my license expired long ago and I never renewed it.

(1)

ING

Internationale Nederlanden Groep is a global financial institution offering banking and insurance services and was Aletta's employer in The Netherlands and Hong Kong.

Did your business idea with the aerial photos gradually gain altitude?

PETER: Not really, because renting an airplane was too costly.

ALETTA: Peter started talking about a balloon with a remote-controlled camera, but I don't think he really meant it seriously. *(1)*

Nevertheless, the path toward going into business on your own was already becoming visible on the horizon.

ALETTA: Well, yes, I suppose so. Peter was working for Philips and I was busy with internships in several law offices. When I began at the ING Group, we both soon realized that we didn't want to end up as employees of big companies, not even if we were ultimately promoted to managerial posts as executives. *(2)*

PETER: Some of the structures were terribly rigid and that didn't satisfy us. One is and ultimately one always remains just another number among many others.

ALETTA: Exactly. But I must also say that I would probably never have taken the daring step without a life partner like Peter. One can never know for sure, but I believe his impulses were very decisive for me. More than once he would say that we ought to dare something. Looking back at our first encounters, I'm not sure whether we had already begun to talk about starting our own business.

PETER: But it didn't take long before our conversations repeatedly centered on the idea of independent self-employment.

So it didn't happen quite so speedily after all?

PETER: Naturally not. Philips had sent me to Hong Kong to develop the areas of consumer electronics and accessories...

ALETTA: ... and I had the opportunity to continue working for ING in Hong Kong, but only in the relatively dull insurance division. My heart wasn't in it, but I accepted the offer anyway because I definitely wanted to accompany Peter to Asia.

PETER: Aletta's frustration was impossible to overlook.

ALETTA: There were 30 people at ING when I started in Hong Kong. All were Chinese and half of them couldn't speak English. It was a real challenge. I learned some worthwhile things, also about Chinese culture and mentality, so my time there wasn't wasted. But ultimately it was too uninteresting and couldn't satisfy my ambitions.

What were the differences compared to Holland?

ALETTA: At home, we would take a coffee break now and then, but there was no such thing in Hong Kong, where they never stopped working. I didn't mind that, because my father had taught me how to work hard. It's in my blood, so to speak. There was plenty of work to do in Holland, but the atmosphere was somehow more relaxed there. And we were also very successful. In Hong Kong, on the other hand, there was a certain doggedness about it.

What were things like at Philips in Hong Kong?

PETER: Awful at first. I was 28, freshly arrived in Asia and assigned to the Kwai Hing district, where Philips had a subsidiary. My windowless office was on the 17th floor, directly behind the elevator.

ALETTA: Whenever I visited Peter, the air stank of oil. And everyone who rode the elevator to the 17th floor stepped directly into his office.

(1)

(2)

PHILIPS

Royal Philips Electronics, commonly known as Philips is a Dutch multinational electronics company and was Peter's employer.

(1)

Peter was responsible for Philips Tapes & Accessories Asia-Pacific.

(2)

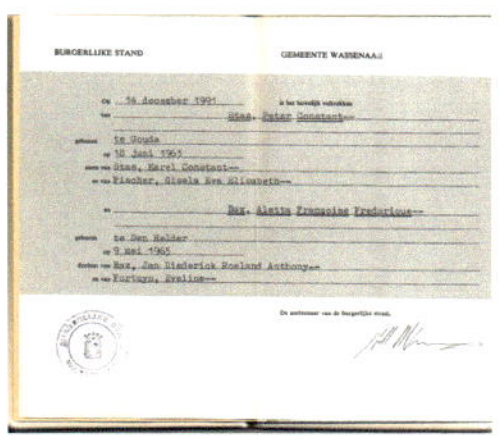

Our marriage certificate

(3)

Philips Tapes & Accessories designed, developed, producec and marketed headphones.

What was your task?

PETER: I was responsible for accessories – audio tape cassettes, to be precise. Philips had a factory in Vienna, but production there was too costly for Asia, so Philips wanted to buy the tapes in Korea and then process them into cassettes in Shenzhen. My responsibilities also involved purchasing the packaging materials and overseeing sales of completed cassettes. *(1)*

ALETTA: Peter was totally on his own when we arrived in Hong Kong.

PETER: On the one hand, that was terrible, but it was also a terrific challenge for me.

ALETTA: Sometimes I would see him only rarely, because he had to travel so much to coordinate everything...

PETER: ... and to negotiate prices, for example, with Lucky Goldstar, LG and SKC. We had to haggle over every last cent. We had to be careful not to buy too much because prices were constantly falling. And we always had to wait until the very last second so as not to miss an even better offer.

ALETTA: The positive aspect was that Peter no longer had a boss directly above him to put the brakes on his ambition. His superior was far away in Vienna.

PETER: I was distantly acquainted with him from my student days and I had sent him my plan.

ALETTA: Weeks went by without an answer, so Peter simply flew to Vienna to discuss everything with him there.

PETER: They told me to take things slowly. They said that there's no hurry. But they also let me do what I wanted. After just one year, we were no longer losing money – despite my good salary and my costly expenses in Hong Kong.

ALETTA: Peter traveled very frequently and I was home alone much of the time.

Were you already married then?

ALETTA: We tied the knot for life three months after our move to the Far East. I really wished that Peter and I could have spent more time together. *(2)*

PETER: If I wanted to succeed, there was no alternative but to travel. In Eindhoven too, where the executives were preoccupied with their own work, they soon realized that plenty of positive things had happened in Hong Kong during my first year there.

ALETTA: Sometimes I had the impression that political intrigues had a much higher priority at Philips than the actual business divisions.

PETER: That doesn't matter. Anyway, one day we asked ourselves, "Why not also include other accessories, for example, microphones and headphones, in this business division?" *(3)*

ALETTA: Fortunately! Philips' CEO in Hong Kong felt that for reasons of prestige, Peter ought to be given a better office. His new office on the 26th floor of Philips' headquarters in Wan Chai commanded a gorgeous view.

PETER: I moved there relatively quickly. We soon grew to become a team of nine. After four years, we were already generating a turnover off 36 million guilders. We earned an uncommonly high profit of eight million guilders.

ALETTA: That's when the problems started.

PETER: You're right. Someone felt that Mr. Stas shouldn't be simultaneously responsible for purchasing and sales. But precisely that was the advantage: I knew what customers wanted at what price and I could pressure suppliers accordingly. And I could also negotiate the numbers of units.

ALETTA: I think one difficulty was that Peter was still quite young. His youthfulness didn't fit in a business with rigidly entrenched structures.

PETER: Yes, I see it that way too. So they promoted me out of the division and made me responsible for new business. An older colleague arrived from Eindhoven to manage the purchasing. From that moment on, things didn't run quite as brilliantly as they had before. Good people left the organization.

What do you mean by "new business"?

PETER: That included computers and televisions which could be connected to computers. Those were rather complex products and they depended on a comprehensive strategy. My responsibility was to develop product concepts and submit them to Eindhoven. Then we were supposed to place and sell the products that had been produced expressly for the Asian market. *(1)*

ALETTA: Philips' internal political aspects played a major role here too.

PETER: Here's one example: I had created an all-in-one computer. There were the most widely diverse responsibilities and competencies. In a nutshell, the business ran rather sluggishly.

ALETTA: That was 1995. I was already busily expanding our watch business. Peter had designed products and accompanied the entire project, but daily business activities were in my hands. In this context, it was necessary to visit customers from time to time. But that wasn't possible for me in Japan because the Japanese wouldn't accept a woman. So Peter had no other choice: every six months, he would take Friday off. After work on Thursday evening, we would fly with a case full of watches to Tokyo or Osaka, where we had big customers. He usually returned with orders for a few thousand watches worth several hundred thousand guilders.

PETER: Every Monday morning there would be a department meeting at Philips. I was 32 at the time and the youngest person at the conference table. My colleagues were in their forties and my boss, the CEO of Philips Consumer Electronics Asia, was about 55. Basically, all one had to do was listen to how the in-house political systems function at Philips: who said what to whom and who had received a promotion. Discussions were often loud and sometimes quite heated. There was almost no talk about products. None other than this boss, who had lived in Japan for eight years, emphatically stated that Europeans would never be able to sell successfully in Japan because the Japanese simply don't accept us and our products. I had to bite my lip to suppress a grin, especially on a Monday morning when I had just returned from a successful trip to Japan. Needless to say, I couldn't contradict him by saying that success was indeed possible.

ALETTA: In 1997, they were discussing the idea of closing the division in Hong Kong and relocating it in Singapore. Peter had been with Philips for eight years and this was definitely the last straw. Even a full year after he had submitted his resignation, he would still sometimes dream about Philips. That's how profoundly it had affected him.

PETER: And I wasn't so deeply involved. I had other colleagues for whom Philips was the be-all and end-all of their existence. Their lives revolved around the company.

ALETTA: That's why on some weekends Peter would ask himself whether he really wanted to spend the rest of his life working for Philips. Time and again, we would consider the idea of building up a business of our own.

(1)

Philips Multimedia Televisions could be connected to PC's for large screen presentations.

PETER: We had the background. After studying economics, my first job was with a consulting firm. But it became increasingly clear to me that I would rather work for a Dutch company which develops and manufactures products. I wasn't interested in petroleum, foodstuffs or cleansing agents, so that ruled out Shell and Unilever. Philips was a good alternative: it made televisions, audio, lighting products and household appliances, it had a staff of 170,000 and it enjoyed a good reputation. That seemed promising. With my plans for an international career, they wanted me to start immediately.

ALETTA: Peter enjoyed his first years at Philips. The opportunities were tremendous. Expanding a division in Hong Kong enabled him to gain a treasure trove of experience and know-how.

PETER: Negotiating with suppliers, developing and selling products, searching for technologies and solutions to problems: that's exactly what we do now, day in and day out, at Frédérique Constant.

And that brings us to our main theme. Tell me about the rather modest beginnings which grew into a genuine success story.

PETER: During my time at Philips, I already owned a small business which had been officially established in 1988. It was based on accessible design software for personal computers with the Windows operating system. I didn't write any programs myself. Rather, my activities were built on an existing program called "Micrographics." In the evenings after work, I had taught myself how to use it to design things. It was a hobby and a passion at first, but it was also very important to me.

ALETTA: We saw watches everywhere then. On TV, in magazines, on posters: watches, watches, watches! It was the end of the "Quartz Crisis." Citizen and Seiko predominated with an 80% share of the market in Hong Kong. Swiss brands were also represented, but not so strongly. Then the first Chinese brands appeared with wristwatches that were fitted with movements from Seiko and Citizen.

PETER: Watches have always interested me. When Swatch first came onto the market, I was at Harvard University and I did a case study about watches.

ALETTA: Peter asked me one day why Chinese watches looked so abominably ugly. They were certainly reliable and their designs were modern, but the watches themselves simply weren't attractive.

PETER: Every time I glanced at my classical watch from Bucherer, I saw a paradigmatic example of a timepiece with an appealing exterior. From that moment on, I devoted greater attention to the theme. I conducted various investigations here in Hong Kong on Saturday mornings. First, I wanted to know where the cases, dials and hands were made.

Where did your inspirations come from?

PETER: We bought old books in which we discovered watches from the 1930s, 40s and 50s. Old catalogues, old advertisements. Those watches appealed to us and lent wings to our imagination. (1)

ALETTA: Then we spent our vacation traveling through Switzerland. We got lists of potential manufacturers from the Fédération Horlogère and we visited about ten of them. Peter had designed the first dial with the aid of Micrographics Designer, so we had something tangible in our hands when we visited potential suppliers.

(1)

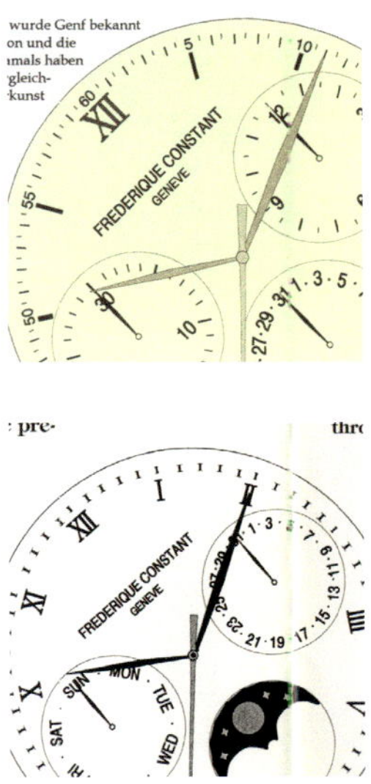

First dial designs created on Micrographics Designer.

PETER: Those sketches proved that our enquiry was indeed to be taken seriously. The Quartz Crisis was coming to an end and suppliers were eager to talk business. We were young and had just returned from the Far East. The manufacturers knew that that's where the future lay.

ALETTA: The Swiss watchmaking industry was gradually getting back on its feet after the crisis, but business hadn't yet really begun to run smoothly. That was advantageous for us because the manufacturers needed new customers.

PETER: The situation would be much more difficult if we were facing the same challenge today. But back then, the doors were wide open and we were welcomed with genuine eagerness to help us, also at the Swatch Group, i.e. at ETA, where we went to inquire about movements. We were given all the documentation we requested without the slightest problem. That gave us all the information we needed about dials, hands and the proper dimensions for fitting into cases. We learned a great deal from that.

Are there any early suppliers who remain in your memory?
ALETTA: Monsieur Duplain comes to mind immediately. He's retired now. He opened the door for us. There was his wife, his daughter and, if I recall correctly, seven coworkers. His dog wandered back and forth. The atelier was in his home. It was a special atmosphere. We explained that we had a little watch factory and we needed cases. The family invited us in. They were very kind and courteous. They listened carefully accepted us as customers. A most friendly and hospitable family!

PETER: Do you remember, Aletta? There's an even better example of the openness and friendliness of individual manufacturers in La Chaux-de-Fonds. We had the list from the Fédération Horlogère, but it specified only the names and the métiers, without a word about the price range or other details. One day we visited someone in the center of town. I've unfortunately forgotten his name. We were invited inside. We explained why we had come and they showed us a few cases, but it soon became obvious that we'd come to the wrong place because their product spectrum included only very noble gold cases for highly exclusive brands.

ALETTA: Right. So they gave us the name of another maker outside La Chaux-de-Fonds. They really helped us because they understood that we were looking for something in a particular quality and at a particular price. Their recommendations suited us perfectly. We were delighted because we were still in a phase of trial and error in those days. Their understanding of our needs was enormously helpful. Step by step, we gradually drew closer to our goal.

Did they freely show you everything without imposing any preconditions?
ALETTA: Surprisingly, yes. Many manufacturers were proud of the companies they worked for, but some covered the names when they showed us the items. We experienced widely diverse situations, but the receptions were always open and positive. Sometimes we found ourselves standing open-mouthed, like children in a candy store.

Why did you buy the cases in Switzerland when they were abundantly available in Hong Kong?
PETER: Purchasing them in Switzerland gave us important advantages, namely: knowledge, tradition and contacts. It was extraordinarily helpful to see how and what they make in Switzerland. Everything is finer there and the craftsmanship is superior. Hong Kong had plenty to offer too. You can find innovation in China. But everything is a bit different. In Switzerland, on the other hand, you experience more fascination.

ALETTA: We ordered only samples at first. No one could be sure we would later order serially manufactured cases. Our Swiss suppliers could have demanded prepayment, but they didn't because trusting us was part of their understanding of what it means to do business.

You lived and worked in Hong Kong. What was the situation there?
PETER: Of course, we diligently made relevant contacts in Hong Kong too – with the Swatch Group, for example, which had an office there that also housed Estima, which manufactured hands for watches. For the dials, we turned to Bock & Schupp, which has its headquarters in Pforzheim. We naturally also contacted Far Eastern producers.

So you gradually grew into a new branch of industry?
Our suppliers' know-how was very helpful. Bock & Schupp had made terrific dials in the past. Viewing and admiring them always brought us one step farther along. One cannot make everything completely

FIRST COLLECTION 1992
2-Hand FC-200E35
Chronograph FC-290E36,
Moonphase FC-260E35,

on one's own from the very beginning. Other people's experiences are extremely valuable on the road toward running one's own business. Good suggestions are the alpha and the omega.

At some point, the components you'd ordered arrived in Hong Kong.

ALETTA: Where we were eagerly awaiting their arrival. After the items from the various suppliers had arrived and as soon as our time permitted, we assembled the first prototypes. All this took time, of course, because we did it alongside our day jobs.

How long did this first phase last?

PETER: A long time. Almost too long!

ALETTA: I would say, roughly 1½ or two years for a total of three models. Then all the parts were available...

PETER: ... and we glued them together with adhesive. Only three of the first six prototypes were functional in the end. Close scrutiny would reveal remnants of adhesive on some of them. But all in all, we could proudly show our little series.

Which movements were fitted?

PETER: Only quartz movements from ETA and Ronda. After all, both Swiss firms operated subsidiaries in Hong Kong.

What kind of watches were those six prototypes?

ALETTA: Mostly two-handed models. One with three hands and one with a date display. Another showed the moon's phases. Still another displayed the moonphase and the date. And finally there was a chronograph.

What was your next step?

ALETTA: Peter took three days off from work. I was still employed at ING, so I likewise had to take a few vacation days. We rented a tiny stand, just three meters by three meters, at the Hong Kong Watch and Jewelry Fair in Wan Chai. Peter, I and a girlfriend of mine staffed it.

Weren't you afraid your employers might discover you there?

PETER: No, because we really hadn't done anything forbidden. But no one from our day jobs came by, although Philips, as a technological company, ought to have shown some interest in this métier. That was another example of how the people there were slowly stewing in their own juices.

Were you successful?

ALETTA: We had almost abandoned the hope of selling anything until at more or less the last minute, a Japanese wholesaler stopped and began looking at our merchandise. He examined our watches, asked about the price, laughed – and ordered 300 pieces!

Before that happy moment, Peter, were you disappointed by the fair?

PETER: Yes, I suppose I was a bit disappointed. But it wasn't that no one was interested in our watches. People came, looked at our products and asked us questions, but no one decided to actually sign a deal with us.

In retrospect, what do you suppose was the reason for that?

ALETTA: Perhaps because we had plunged into the adventure without a business plan. We hadn't defined the goals we wanted to achieve in a specific timeframe. We were earning salaries at our day jobs so we weren't dependent on achieving any specific successes at the watch fair.

PETER: From that viewpoint, we weren't thinking about any conditions at all. We hadn't gone to banks to borrow money that we would have to repay. We weren't burdened by financial obligations. We simply showed our products at the fair and waited, not entirely without nervousness, to see what would happen.

What would you have done if the fair had ended without a sale?

ALETTA: We probably would have been very disappointed. Perhaps we would have abandoned the project altogether. Who knows? Fortunately, the question never arose. We were simply satisfied with what we'd achieved.

That no doubt also raised the question about continuing your employment at ING.

ALETTA: When we had the contract for 300 watches in our pocket, so to speak, there was no further discussion about my future. I had no choice but to quit my job at ING to handle that first big order.

When you left ING, was there a tear or a twinkle in your eye?

ALETTA: I left with a smile because I didn't really enjoy my job in Hong Kong. As I said, ING had only its insurance division there and I had accepted the offer solely because I wanted to go to Asia with Peter. My heart wasn't in the insurance business.

Let's talk about money and financing. Had you saved money to finance your entry into the watch business?

PETER: Not really. That's why we asked our customers to pay 30% in advance. The remainder was payable before the merchandise was shipped.

ALETTA: We always spoke candidly with suppliers and customers. We would explain that our business was still very small, we financed everything on our own and we needed a prepayment or a security guarantee before we could begin production.

Your candidness was greeted with acceptance?

ALETTA: Yes, our partners were very understanding. Prepayments weren't a big problem. Suppliers also trusted us completely, so we were able to cultivate excellent business relationships in both directions. There's always a risk when you work with someone you don't know.

So a new phase in the life of the Stas Family began with your success at the fair?
PETER: You could say that.

ALETTA: After quitting my job at ING, I was my own boss and totally independent. I could negotiate much better conditions with potential suppliers. We also had to write orders to ETA, Ronda and Estima for hands. Everything ran from Hong Kong.

PETER: We arranged to manufacture 350 timepieces, although only 300 had been ordered. That way we would have samples for the future and a few watches in reserve which we later sold to Taiwan. Aletta monitored the progress on a daily basis.

Were you satisfied with the first products?
ALETTA: Considering the circumstances, yes. If we had looked at their quality from today's vantage point, we would perhaps have been shocked. But back then we didn't know anything else. The quality was perhaps only marginally acceptable, but what more could we hope to achieve with our tiny budget? The price was so low. We ordered cases plated with a layer of gold that was five microns thick. So the cost-benefit ratio was satisfactory.

What about your Japanese customer?
ALETTA: He came to Hong Kong, paid and took the watches with him. We had financed all 300 of them without outside help. Thirty percent were prepaid. Then there was our own money that we had invested. Our budget was scanty, but it sufficed.

PETER: Those 300 watches had cost circa 100,000 Hong Kong dollars. The unit price was between 40 and 60 Swiss francs. One mustn't forget that those were quartz watches and their cases were plated with a five-micron-thick layer of gold. These early timepieces were really only a first step into a field of business that was still totally unknown to us. Everything was based more or less on trial and error because we had no firsthand experience whatsoever.

ALETTA: Some of the tolerances weren't quite right in those watches, but they were part of a great learning process.

Did those watches already bear an insignia?
PETER: Yes, they were signed "Frédérique Constant."(1)

How did you come up with that name?
ALETTA: Every child needs a name, so we often discussed the topic. There were two options: we could follow in the footsteps of Patek Philippe or Vacheron Constantin, i.e. we could choose a French name that would give us a more luxurious image...

(1)

FREDERIQUE CONSTANT
GENEVE

Frédérique Constant logo w th family crest and Geneve

FREDERIQUE CONSTANT
Brand name based on great grand parents Frédérique Schreiner and Constant Stas

PETER: ... or we could opt for a short and catchy name like "Rolex." Our family name "Stas" would have sufficed, but an "S" at the beginning and another "S" at the end didn't look pretty and we didn't want to start the business with our family name.

ALETTA: We began to think about an authentic French name with a history. Our great-grandparents' middle names occurred to us: "Frederique" came from my family and Peter contributed "Constant." We were very pleased with the result: "Frédérique Constant." Peter's great-grandfather Constant had worked for an etching and printing factory that also made dials for Dutch domestic clocks, so there was a real connection there. One of my ancestors was named "Frédérique." The names were waiting for us.

How long did the name search take?
PETER: Finding the names per se went rather quickly, but afterwards we were doubtful and weren't sure if we'd made the right choice. First, because of our own names and also because Frédérique is a woman's name. When we had overcome our last shred of doubt, we arranged to have the name "Frédérique Constant" registered as a protected trademark in Switzerland.

Does the feminine element in your company's name have a negative effect? A woman's name is quite unconventional in the world of watchmaking and is probably absolutely unique there.
ALETTA: Only people who speak French notice that "Frédérique" is a woman's name. And especially in France, our successes speak for themselves.

Frédérique Constant also has a logo.
PETER: Yes, the coat-of-arms of the Stas family.

Let's get back to your first watches. Was your customer satisfied with the delivery?
PETER: Absolutely. He was very pleased and soon returned with an order for 1,100 more.

What sort of wristbands kept them affixed to their wearers' wrists?
PETER: Our first straps were made of leather.

ALETTA: Later we had a few models with metal bracelets. We learned the hard way because we had initially worked in an uncoordinated way. We should have given the same sketches to both the case-makers and the bracelet-makers. That would have kept production

SLIM LINE
FC-235M36 & FC-235M26
(1995)

closely interlinked. Our failure to do so meant that the first bracelets didn't quite fit their intended watches. We had to take the bracelets into our own hands and post-process them to complete the wristwatches.

PETER: A mishap like that happens only once. We learned our lesson!

Did anyone complain about the quality of the cases?
PETER: No. The watches were inexpensive and expectations were not exceptionally high. Later we occasionally encountered a few small problems, but they're really not worth mentioning and we didn't lose any customers because of them. I manually assembled the first watches myself, but later we needed professional help.

ALETTA: ... which we found in Hong Kong. In those days, watches that had been assembled in Hong Kong could still bear the words "Swiss made" on their dials.

PETER: It depended on the components. So we collaborated with a Swiss citizen in Hong Kong who also worked for other Swiss brands. But we soon found an assembly factory in Switzerland.

ALETTA: A skiing vacation in Switzerland was coming up, so we revisited our suppliers. We found Philippe Bouchet in Geneva. He had a tiny office on one floor of a building and was willing to organize the assembly of our watches. (1)

PETER: Our next series was assembled here in Geneva, in an atelier with a staff that also worked for other brands, including some highly renowned brands.

ALETTA: The finished watches were shipped to us in the Far East, where we attached the wristbands. Needless to say, we delivered everything in beautiful etuis.

What else remains to be said about that second series?
PETER: We made small modifications, including a new Breguet-style dial from Bock & Schupp, but basically they were the same watches. This was also when I first met Mr. Miguel Garcia, who was sales manager for Sellita at the time. He was in Hong Kong for the watch fair and he hosted a dinner. It had been recommended to us that we attend, so we gladly accepted his invitation.

ALETTA: In those days, various suppliers advertised for one another by word of mouth. It was very nice. They would say, "Go here" or "Go there, you'll find the help you need."

PETER: Mr. Garcia suggested we use self-winding caliber 2846 in our mechanical watches. It's a relatively thick movement with a balance that oscillates at a frequency of three hertz. Nonetheless, it was rather costly for us. Then, if not before, we had to face the question: "How do we want to proceed?" (2)

ALETTA: We discussed it with Mr. Garcia and we realized that the plate of the 2846 could be pierced to make the balance visible, thus showing that it is indeed a mechanical movement.

PETER: No sooner said than done. Sellita sent us a few movements with cut-out plates. We made the dials for those watches ourselves. The apertures weren't symmetrical, which didn't look very good.

(1)

FIRST OFFICE
Frédérique Constant in Troinex, Geneve

(2)

FIRST AUTOMATIC
Mechanical self-winding caliber 2836

MANUFACTURE
Frédérique Constant Heart Beat Manufacture FC-910MC3H6 (2004)

ORIGINAL
Frédérique Constant Heart Beat FC-310M36 (1994)

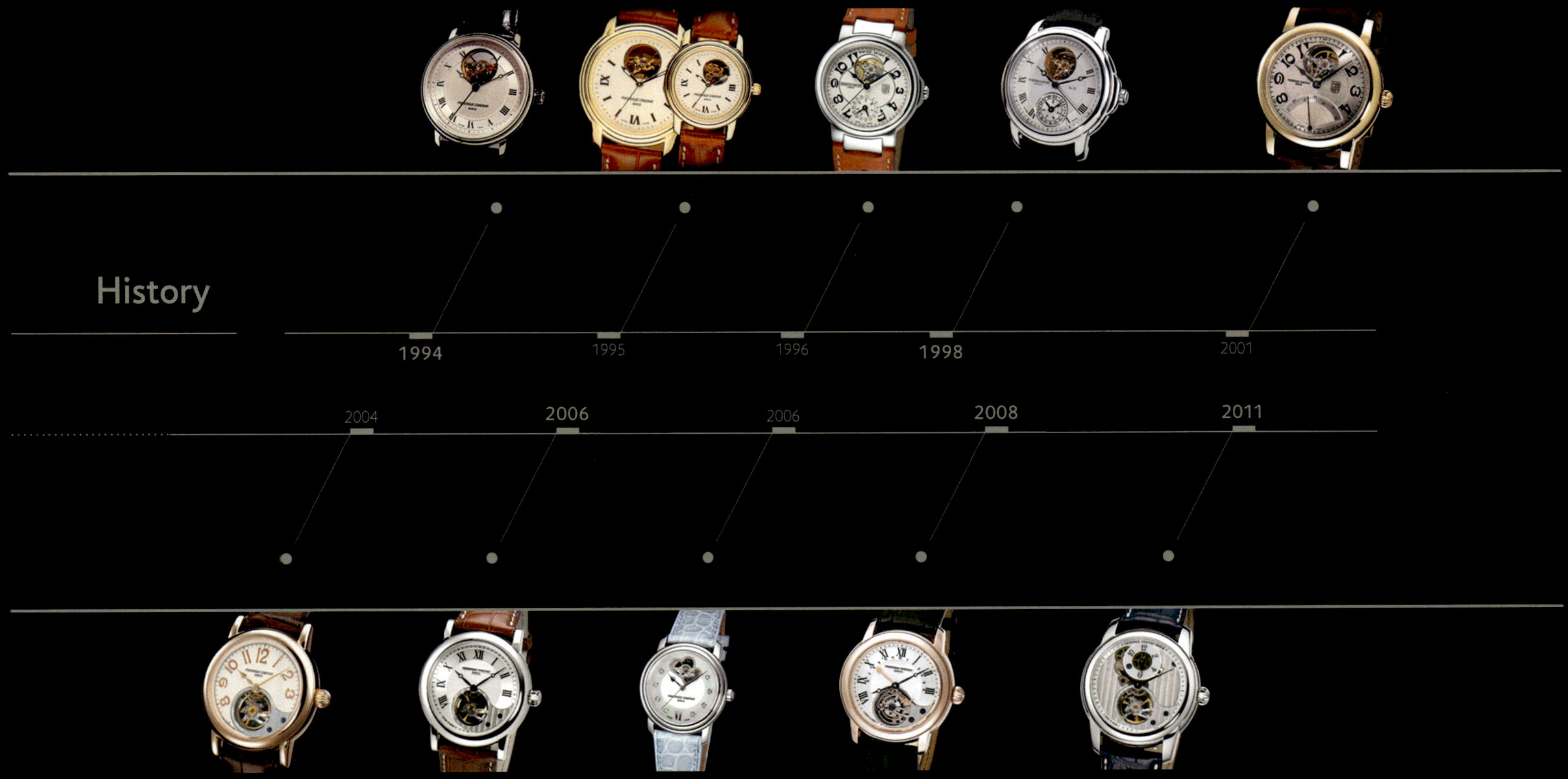

CHRONOLOGY

- **1994: Original Heart Beat introduction**
- 1995: Lady Heart Beat introduction • 1996: Highlife collection • **1998: Heart Beat Day Date**
- 2001: Heart Beat Retrograde • **2004: Heart Beat Manufacture**
- **2006: Heart Beat Manufacture Automatic** • 2006: Ladies Automatic Double Heart Beat collection
- **2008: Tourbillon Manufacture Silicium** • 2011: Heart Beat Manufacture Dual Time

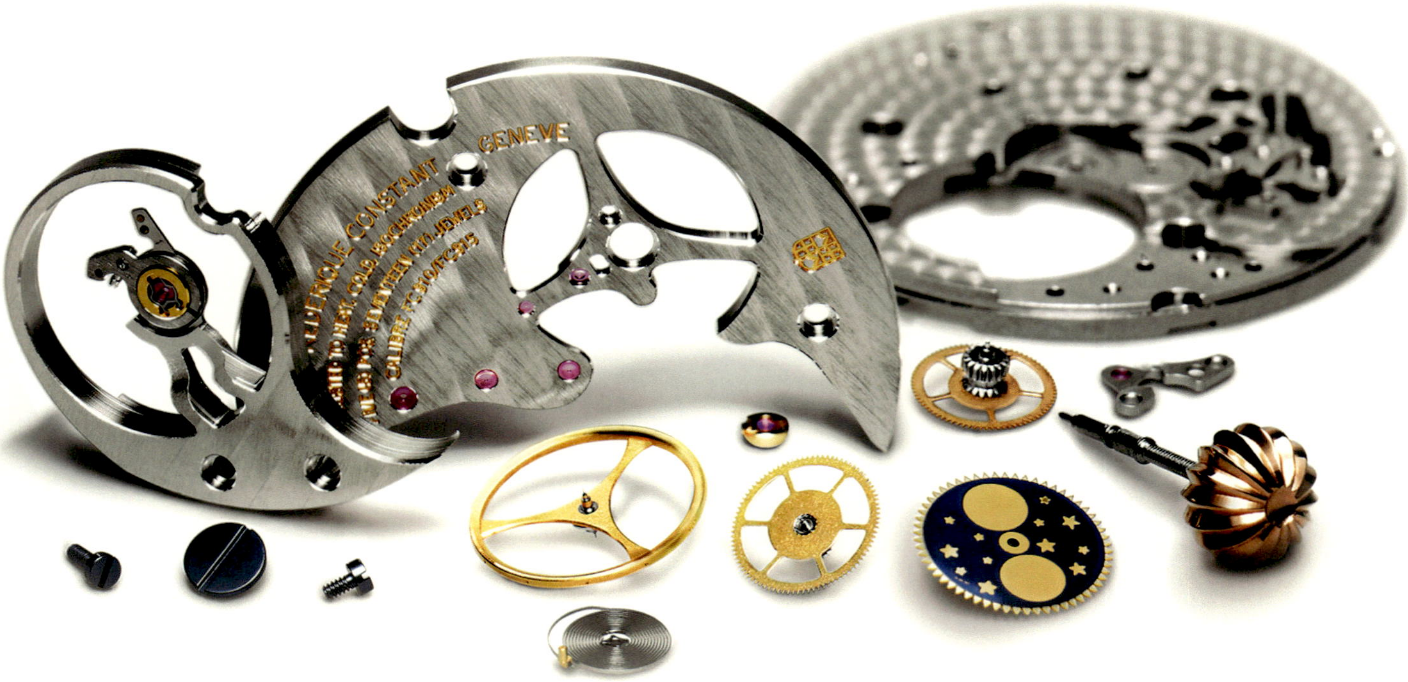

HEART BEAT MANUFACTURE
Parts of FC-910 Caliber (2004)

(1)

Heart Beat opening

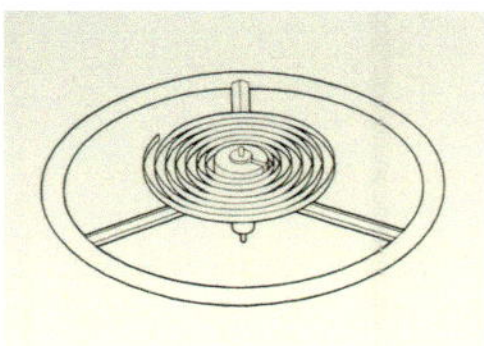

Balance Wheel & Spiral

But you persevered?

PETER: Yes, because we were convinced by the idea. Unlike our competitors who copied us later, we had the plates opened even more, so the look was right and the aperture was shifted into the middle. That's how the "Heart Beat" story began. (1)

What happened afterwards?

PETER: I designed models for the next series on my computer in Hong Kong during my free time. Aletta took care of the administrative work in a friend's office that we had sublet.

ALETTA: I phoned suppliers and paid invoices. We also needed a brochure.

PETER: Aletta and one of her girlfriends again rented a stand at the watch fair in Hong Kong to present the new models. We acquired customers from Brazil, as well as Becker from Hamburg in Germany.

ALETTA: Yes, Becker Jewelers, Nicolas Giercke came on board early on. A Dutch department store also purchased our watches.

PETER: Our early customers accepted a big risk. We were totally unknown, but they agreed to prepay 30% of the purchase price.

ALETTA: The arrangement was a bit different with Becker, to whom we sent twenty watches with an invoice payable after delivery.

PETER: We also made private labels during that period, which enabled us to achieve larger quantities. For example, we made watches for a customer in the USA who provided the desired designs and benefitted from our close contacts with manufacturers. We also made promotional watches for Heineken beer and the Volvo Ocean Race.

ALETTA: We delivered watches for the Brazilian market without the Frédérique Constant insignia. The order had been placed by a large retail trading group.

PETER: We couldn't afford to be too choosy in those early days. We were glad for every order and we accepted everything that seemed likely to be at all profitable. That was the only way for us to become established. To return to the financing for a moment: our own money wasn't enough to finance production when the order for 1,100 watches arrived, so a friend agreed to loan us 60,000 guilders.

That was "risk capital," so to speak?
PETER: Definitely. But it paid off for our friend. He gave us the money in 1992 and in 1997 he got back circa 300,000 guilders.

ALETTA: Another friend invested 70,000 guilders in shares of the company two years afterwards. We repurchased half of those shares a year later and bought back the remainder two or three years after that. By 1997, we had repaid all outside capital that had been loaned or invested.

Frédérique Constant is truly a success story with continual growth in the number of watches produced annually.
PETER: Yes, the numbers of units produced have increased annually right from the start. We made 11,000 watches in 1997. That's also remarkable because very dissimilar things occurred during this era.

What, for example?
ALETTA: On the one hand, we had to discontinue collaboration with our assembly partner in Geneva because, among other problems, quality and delivery times were no longer satisfactory. We had nothing but problems. I constantly had to fly back and forth between Hong Kong and Switzerland. We weren't satisfied with that.

PETER: Then Philips decided to relocate my business division to Singapore, which meant that we would have to move. We didn't want to leave Hong Kong because that would have been the end of our watch manufacturing venture, which was already flourishing so well that it would have been a pity to abandon it. So I quit. Philips even offered me a job in Palo Alto, California, but the offer came too late. In my mind, I had already left Philips far behind.

Weren't you nervous?
ALETTA: Of course! Before Peter wrote his resignation, we took another vacation trip. The people at Ronda had often asked us why we didn't come to live in Switzerland. They said, "You make beautiful watches with strong potential and you achieve good numbers of units produced, so you really should settle in Switzerland."

PETER: If it hadn't been for the watches, Palo Alto might have interested me. But Philips had lost its appeal. At 33, I was too young for in-house political intrigues. I had risen through the ranks too quickly. I freely admit that.

How did your revenues look when you resigned from Philips?
PETER: In my view, revenue seemed too low. The total was not quite two million francs when we settled in Switzerland. We had to pay twelve people in Hong Kong along with all our other expenses with that money. The margin was less than 25 percent. It was a big step and careful forethought was essential.

ALETTA: Before moving to Switzerland, we also needed Swiss work permits, which meant that we had to talk with the authorities. They agreed to issue us the permits under the condition that we employ at least twelve people in Switzerland within the next three years.

ALETTA: This meant that the originally envisioned plan, with Peter and me living in Switzerland and our company staying in Hong Kong, would be impossible. Our business would necessarily have to follow us to Switzerland. But in return, we were granted exemption from taxation for the first ten years.

PUBLICITY
Artistic impression in earlier brochure (1994)

POCKET WATCH
FC-435MP5 (1995)

Before we talk about Geneva: Aletta, you frequently played hockey in Hong Kong, didn't you?

ALETTA: Yes, I started playing hockey at age seven. It's a popular sport in Holland. I wouldn't be surprised if half the population of the Netherlands plays it. There's a Dutch community in Hong Kong. There's been a Dutch hockey club there for many years because the Dutch have traditionally been active throughout the world. Especially when they're far from home, they seek out one another and form an expatriate community. So there were two men's hockey teams and one women's team in Hong Kong. We played against Chinese, Australian and, of course, also English teams. Those were social events for us.

How often did you practice?

ALETTA: Once or twice a week, depending on how much time I had available. The games were played on Saturdays.

Other than that, you focused entirely on your watches after you resigned from ING?

ALETTA: Yes, it was sometimes quite stressful at first, especially when we urgently needed to complete something. There was so much to do! We had to collaborate with our Swiss partners. I recall several hectic periods. But there were also quieter weeks. We primarily acquired new customers and new orders at trade fairs, first in Hong Kong and then in Basel starting in 1995. We also traveled to Dubai once with a delegation from Hong Kong. We as Dutch people weren't supposed to be part of the entourage, but everything had already been organized so we simply boarded the plane and flew. We stayed at a lovely hotel in Dubai

When was that?

PETER: 1993 or 1994, if my memory serves me correctly. Local people came and we received several orders. One of them was huge. We agreed to deliver 500 ladies' watches with solid 18 carat gold cases. The order came from the sultan of Oman, whose portrait was to be printed on the dials.

That must have been a major challenge.

ALETTA: Certainly, because we had to buy enough gold to make all those cases. It was an enormous effort on our part.

Did the big order end well?

ALETTA: Fortunately, yes, but we trembled a bit at times. Especially the dials with the sultan's portrait were very demanding. For a short while, we were concerned that the dimensions were wrong, but everything turned out right in the end. We shipped the merchandise and everything was punctually paid for, exactly as we had agreed.

Was this followed by similar orders from the Near East?

PETER: Yes, but in much smaller dimensions. Even today, I still don't know how that deal with the palace came to be. Most nobles in the Mideast patronize famous brands, not unknown newcomers like us. Perhaps it was a spur-of-the-moment decision and our competitors weren't able to deliver in the desired timeframe.

What was printed on the dials of those watches?

ALETTA: "Frédérique Constant," of course. That entailed a certain risk for the sultan's agents. No one knew us. Everything could have gone terribly wrong. We couldn't name any references. Nothing. If we had had an agent in the country, someone who could have put in a good word for us and help to seal the deal, then it would have been much more understandable. But we had nothing of the sort there. The watches were intended for graduates of a nursing school, I believe.

Let's talk about your move to Geneva.

PETER: We had mixed feelings. Aletta, I and our nine-month-old son with trepidations in our luggage. I was leaving a real global player behind and looking ahead at a big personal risk. Mr. Bouchet worked only for himself and didn't really have a business.

ALETTA: Things weren't easy from a financial viewpoint. We made do without relying on banks. We never asked a bank for a loan.

OFFICE
Carouge 1997-1999

PETER: Our first contact with banks came when we were planning to build our own house, but only for a mortgage, not for working capital. Later there was the mortgage for our company's building in Plan-les-Ouates.

Does the Stas Family now have Swiss passports?
PETER: No, we're still Dutch citizens, and that's good.

ALETTA: In the past too, I'd always categorically said "no" to a Swiss passport. In the meantime I've changed my mind and would accept dual citizenship, i.e. Swiss and Dutch. That could be good for business. I would have to research it. But I'm only marginally interested, unlike some of our foreigners acquaintances here, many of whom invest a lot of energy in it. I would never give up my Dutch passport.

Your work permits are permanent now?
PETER: Yes. We were required to report regularly on the progress of our business during the first few years. After temporary "B" permits, we were later granted unlimited "C" permits. *(1)*

How did things develop in Geneva?
PETER: At first, we moved into a small building, but we only stayed there for four months. We moved within Geneva to Carouge in 1997, where we rented one floor with an area of 120 square meters on which we opened

(1)

Permanent Work Permit
Switzerland

(2)

INTERNET
First Frédérique Constant website (1995)

our first own atelier to control the quality of incoming merchandise. Three or four people worked in the office and five were employed in the atelier.

You were relatively unknown newcomers in Geneva. Was it difficult to find watchmakers?
ALETTA: Not necessarily. Several people who had worked for us in the assembly atelier were glad to switch to Frédérique Constant.

How did your brand continue to grow in Geneva?
PETER: Annual production quickly increased from 11,000 to 18,000 watches, most of which fitted quartz movements. The larger numbers of units were mostly private-label products. But this business division soon shrank because we set different priorities and increasingly focused our attention on Frédérique Constant.

ALETTA: Our move to Geneva coincided with a dramatic improvement in our catalogues. We also launched our Internet site. *(2)*

Did Frédérique Constant already offer a regatta watch then?
PETER: Yes. Here's how that came about. My nephew was sailing in the Olympic Games and he asked me for an appropriate watch. Together with Camy, we developed a chip to control the disc for the pre-start phase. That was a complex task and it nearly overwhelmed us. But all in all, it was a success. We manufactured about 3,000 of those watches.

It was originally named "Yacht Master" and it had to be renamed.
ALETTA: A representative from Rolex suddenly came to our stand at Baselworld. He prohibited us from using the name "Yacht Master," to which Rolex held exclusive rights. The panel insisted that we remove the watch from our display window.

PETER: But a few minutes later, I already had the idea for the name "Yacht Timer." Rolex's lawyers couldn't find any reason to prevent us from using that name.

FREDERIQUE CONSTANT
GENEVE
SWISS MADE
FREDERIQUE CONSTANT
GENEVE
RESERVE
DE MARCHE
SWISS MADE

YACHT TIMER
FC-298V3A5 with prop-ietary regatta caliber (1997)

When did Frédérique Constant first participate in Baselworld?

PETER: In 1995. Again, I took a vacation so I could be there in person. Philips was entirely unaware of those activities. They had heard that Aletta was doing something with watches, but they didn't take it seriously. Basically, no one at Philips was interested in anything beyond the limits of Philips' own product spectrum.

Do you still cultivate contacts with Philips?

PETER: I recently ran into the CEO. He and I had lived in the same student house years ago. He's three or four years older than I. We chatted and afterwards he sent me an email congratulating me on my success with Frédérique Constant. Sometimes I wonder if they had even the slightest inkling about my double life. But it really doesn't matter anymore.

CLASSICS
Original Heart Beat FC-310M35 and Power Reserve FC-340M85

Let's get back to Geneva now.

ALETTA: Frédérique Constant was still very much in flux when we arrived in Carouge. We made watches with our own insignia and with those of other labels, and we continued to grow steadily.

PETER: After three years there, our premises were again too cramped. So we moved to Chêne-Bourg for 5½ years, where nearly 800 square meters were available for us not far from Rolex.

ALETTA: We set up an advisory board with Wim van Heijst and Hans van Willige. We also spoke with professional advertising agencies about ways to better evolve our brand.

PETER: When the premises at Chêne-Bourg became too small for our needs, we searched throughout Geneva for a suitable building. But we couldn't get the one we wanted. I never found out why they turned us down.

ALETTA: A real estate broker asked us, "Why don't you consider constructing something of your own." We would then be able to get a plot of land in hereditary leasehold from the foundation for industrial real estate in Plan-les-Ouates. It would cost only 15 Swiss francs per square meter per year. That's a big help for us from the community and the local government.

PETER: Together with an architect, we designed the building to meet our needs: logical, efficient, practical and accessible. We made 48,000 watches each year at that time. The new building dramatically altered our business's profile. Suddenly, we got noticed. And people asked, "Where do they come from?"

Were there problems with financing?

PETER: No, none whatsoever. The bank readily granted us a mortgage. Our business was successful and the location in Plan-les-Ouates was absolutely credible.

ALETTA: Plan-les-Ouates was only our second choice at the beginning. Commuting from our home takes half an hour. It was only a five-minute drive from Chêne-Bourg.

ADVERTISING
Earlier Heart Beat Campaign (1996)

ATELIER & OFFICE
Frédérique Constant
Chêne-Bourg, Geneve 1999 - 2005

PETER: Looking back, we're very happy to be here. Customers and journalists experience us in the immediate vicinity of Piaget, Harry Winston, Vacheron Constantin, Patek Philippe, Rolex and Ralph Lauren. That's ideal and it raises the status of Frédérique Constant.

ALETTA: When we decided on Plan-les-Ouates, we weren't so keenly aware of how advantageous the location would be. Interestingly, our neighbors estimated that the building would cost much more than it actually did.

Why is that?
PETER: It had to do with architects who build much more elaborately and more cost-intensively than ours. I was intimately involved in the new construction.

We commissioned an engineer from East Germany. At first they said, "No, that's impossible," but it ultimately turned out to be possible after all. He had all the necessary qualifications. He brought the concrete skeleton and the façade components from Germany. The steel came from northern Switzerland. Canadians built the ceilings. We personally drew the plans for the furnishings, which we arranged to have manufactured in China. We've been here for six years and we feel very comfortable here. This building suits us quite well.

What prompted you to choose Geneva for your company's headquarters?
ALETTA: It's certainly good to have the phrase "Swiss made" on our dials. But Geneva is even more exclusive, so it further enhances the value of our products. If you want to set yourself apart from your competitors, you've got to do the very best you can. And there are few Genevan brands in our price range, so that gives us a unique selling point.

MANUFACTURE
Frédérique Constant Plan-les-Ouates, Geneve

FREDERIQUE CONSTANT
GENEVE

Did you ever consider the Jura region?
PETER: The thought never crossed our minds. We wanted Geneva because of the signature. And then we had to decide on a residency location in Switzerland. Geneva suited us perfectly.

ALETTA: Philips wanted Peter to relocate to Singapore and our business continued to grow, so we had to take action. Our watch assembly was based in Geneva and the manufacturing process needed someone to monitor it. After lengthy consideration, we decided to move here.

Were there moments of anxiety when you doubted whether everything would work out and you were uncertain if long-term success would ensue?
ALETTA: When I look back, I can't remember everything. I'm blessed with the enviable ability to recall only pleasant things. I can easily forget the bad ones. So I can say now that I wasn't anxious or worried, although I'm more likely to be the hesitant one. I consider all the risks and difficulties that could arise. I didn't see any insurmountable problems, but I wasn't sure that I wanted to live in Geneva. To work in Switzerland would be one thing, but to actually live there? I had lived here once in my childhood. My uncertainties were more related to private life than to the business aspect. Perhaps I was being naïve.

PETER: Aletta said, "Let's give it a try for two years. If it doesn't work out, we'll make a new decision. Whatever happens, we'll be able to add a Swiss adventure to our résumés. That can't hurt and it's likely to be helpful if and when we ever have to search for new employment."

Have you ever regretted your decision?
ALETTA: As far as the work is concerned, no. But to be perfectly honest, I did initially regret it from the social point of view. The difficulty of genuinely participating in social life here caused me certain problems. But in the meantime, I'm very happy here. We really feel comfortable. I've changed, and now I'm better able to accept traditional customs.

You run a genuine family business. What advantages does that have from the owners' viewpoint?
PETER: Aletta and I are responsible for our business. We've cultivated and evolved our idea of "accessible luxury" for the past twenty years. Rather than looking nervously toward the right and the left and allowing competitors to distract us from our chosen path, we've logically and self-consistently expanded our collection.

ALETTA: Compared to the big bulldozers, everything runs with a more personal touch. That's obvious in the sales and in collaboration with agents. When a big corporation takes over a watch brand, longstanding contracts are all too often rashly cancelled so the corporation itself can handle distribution. That was good for us because we found good distributors who were open for a new brand. We also feel the advantages at independent specialized dealers because many of them prefer to work with families rather than with big groups that often exert overwhelming pressure. If a jeweler wants to sign a new contract with a strong-selling brand or renew an existing agreement with a popular label, he's often compelled to also accept more difficult brands. Furthermore, he's obliged to achieve specified revenue targets. Things are easier and there's more understanding when he collaborates with a family business.

Do you both enjoy making decisions?
ALETTA: Yes, that too might be one of the advantages of our business. We decide speedily and that stimulates our employees to do likewise. One knows one's situation quickly here. One generally sees the results of a decision very soon. Good ideas are put into practice without delay.

PETER: That's quite unlike the situation at big corporations, where decision-making pathways are typically very lengthy, complicated and characterized by what you might call a "limited-responsibility mentality." The speed and directness can be frustrating as well for our colleagues here because some things simply aren't possible.

How does the outside world perceive the Frédérique Constant watch manufacture?
PETER: We have a very good reputation. Many customers initially don't understand that very costly luxury brands can share a common roof with very simple entry-level brands. For example, when collectors learn that, some of them are very surprised. Many of them are familiar with this situation, but there are also quite a few watch aficionados who have never heard of such contexts.

ALETTA: The art of watchmaking still has a craftsmanly touch. I believe a family business offers an additional advantage here. Many watch lovers find it more charming to own a timepiece from a family-run manufacture.

HEART BEAT PAIR
FC-310M36 & FC-310M16
(1994)

Could it be that some jewelers, which operate as family-owned businesses, would rather cooperate with a family-run manufacture than with brands that belong to big corporations?
PETER: Yes, definitely. It's a totally different kind of partnership.

ALETTA: They obviously like it. But I've been doing less traveling of late, so I see this more from a distance and my opinions are mostly based on what Peter tells me when he returns from his trips. Nonetheless, what I've seen on my own trips, which are comparatively few, confirms this belief. The family spirit is wholeheartedly welcomed.

How often do you travel for your business?
PETER: I'm usually away one week per month. Of course, I use the opportunities to visit the shops. Many conclusions can be drawn from what I see there: for example, what we need to improve in our marketing, in our advertising, etc. Beyond that, I get plenty of ideas for new watches. I also meet with many journalists while I'm traveling. They're very helpful for raising brand awareness in the various countries.

ALETTA: Naturally, we cultivate contacts with our customers at Baselworld. We have circa 2,700 customers worldwide and we see about 400 of them during the fair.

Your agents have plenty of requests and suggestions for new models.
PETER: Yes, of course, and we make a note of every idea. If it fits our concept, our strategy and our collection, we'll go ahead and create prototypes, which we then show to our distributors to learn their opinions. Afterwards, about 15 to 20% of all the models we've shown them undergo further modifications, which continue until the last minute. In the most extreme instances, we've cancelled an entire line.

Why did you do that?
PETER: Our distributors felt that the dials were too different and no longer reflected Frédérique Constant's identity. Their opinions convinced us and we abandoned the project, albeit with a heavy heart.

ALETTA: Fortunately, we hadn't yet manufactured a huge number of watches. But twelve models lay on the table. We had worked very hard on them, but we sadly had to eliminate them in their present form.

PETER: After hearing the justified criticisms, we reworked existing dial designs so they would suit the new models – and then the line was a real success.

How could something like that happen at Frédérique Constant?
PETER: It's very simple. We had briefed an extramural designer clearly about the creation of the new line, but her final results were too different. We ourselves could have lived with them more or less, but our customers weren't satisfied. They were strongly opposed to the new line because it lacked brand recognizability.

Let's talk about the evolution of Frédérique Constant's collection from your viewpoints.

ALETTA: We began with six models for men. Soon afterwards we also created ladies' watches as smaller versions of our gent's models. But watches for women never generated more than 10% of our revenue, perhaps even less than that, because these watches really didn't have a special image. We found many inspirations, mostly for gent's watches, in the old books. Women have always been treated like neglected stepdaughters in horological history. I can more readily feel what men like. The traditions are more clearly perceptible.

Younger men are increasingly fashion-oriented nowadays.

ALETTA: Yes, the tides have changed somewhat. Nevertheless, and analogously to our watches, it's much easier to design a classical wardrobe for men than for women. We looked at what the famous watch brands had made for women through the years and we didn't discover anything genuinely stimulating. That certainly wasn't where we wanted to go with Frédérique Constant.

PETER: Eccentricities aren't appropriate for our brand. Furthermore, risk grows exponentially when a product becomes more extreme. We always want to feel comfortable with what we make. We would have felt uncomfortable with creations like those, so we initially restricted ourselves to smaller models for men.

One should never underestimate feminine hedonism!

ALETTA: Yes, at some point, we realized that women are becoming increasingly more independent and that they use some of the money they've earned to buy luxurious items. In the past, it had been very unusual for ladies to buy their precious rings with their own money. Nowadays this happens much more frequently. Times have changed and we had to respond to those changes.

PETER: We saw opportunities for expansion in that direction. Early on, we included in our collection the "Heart Beat," with its balance visible from the front. Then we had the smaller version for ladies. When we read all the articles published about it, we definitely wanted to bring something more feminine into the collection.

Diamonds or mother-of-pearl would be a likely choice.

ALETTA: Yes, that's exactly what we did in our classical "Heart Beat" with its circular window at the "12." We added diamonds to the dial. And they sparked gratifyingly strong interest.

PETER: But we had to think of something new when we saw that we were being copied increasingly more frequently, not only by cheaper brands, but also by better-known and prestigious brands. We recalled the heart-shaped aperture that we had discussed with our Asian distributors several years previously.

DOUBLE HEART BEAT
Iconic world
best-seller for Women

ALETTA: A simple heart seemed too simple to us, so we decided to make two hearts. They represent love, the relationship between a woman and a man or between a mother and her child. Hearts are appealing and they make it different.

I'm sure you secured legal protection for the two hearts.
PETER: Yes, because we needed to protect our uniqueness.

Did success arrive quickly?
ALETTA: Yes, you could say that. But like everything else in our company, this too progressed step by step. We continually optimized the product and its advertising. That's part of our philosophy. The sum of all these efforts made this wristwatch, i.e. the one with the chocolate-brown dial, a tremendous success. It's one of our three bestselling items. And that includes the men's models. But this succeeds only when one makes the right decisions and takes the right actions in all areas.

What does that mean specifically?
PETER: Design, quality, advertising, marketing....

Which is your bestselling item?
PETER: One must distinguish between revenue and numbers of units, but I would certainly mention the "Runabout Automatic," which has replaced our classical self-winding model. We have only limited numbers of units for the Runabout and that's why the bestseller changes among various models in this line.

You're personally involved in the creation of ladies' models, aren't you, Aletta? It's well known that women know best what women want.
ALETTA: Yes, I'm very much involved there and I have been for many years. Peter does much of the preliminary work on the computer because he's really talented. That's also true with the men's watches. Sometimes I don't have enough time or energy for the creative aspect because I have so many other things to take care of.

All your watches reflect your philosophy of accessible luxury. How did you come up with the idea?
PETER: It arose together with our lives and the evolution of our careers. At first, we didn't have enough money to buy costly watches. Precisely there is where we saw the niche for our startup. Beautiful wristwatches for accessible prices were very rare. Successes in subsequent years proved that our idea was right. So we consistently continued and we expanded the Frédérique Constant brand in that direction.

How much money are your potential customers willing and able to spend for a wristwatch?
ALETTA: A retail price of 1,000 euros is already a considerable investment in the watch business. There are renowned and traditional brands that offer Swiss-made watches for 200 to 300 euros. Basically, no one really needs a wristwatch that costs more than 100 Swiss francs. It does what one expects it to do.

Recent years have seen watch prices increase much more steeply than incomes have risen. As the owners of a family business, how are you able to develop and market accessible priced products that look much more costly than their actual prices?
PETER: We have to calculate very carefully. Sometimes we make do without a margin. And we must manufacture with the utmost efficiency. Our information technology must be optimally well developed. We've been very strong in that right from the beginning and it's true again with our iPad Apps. This helps us save money and we can pass those savings along to our customers by selling them our merchandise at attractive wholesale prices. A certain volume is also important: smaller numbers of units are necessarily costlier. What one loses with a narrow margin can be at least partly recovered by manufacturing larger quantities.

page 44-45
JEWELRY COLLECTION

1. 18K Rose Gold Earrings Double Hearts FC-602A9
2. 18K Rose Gold Double Hearts FC-103A9
3. 18K Rose Gold Pendant Double Hearts FC-204A9
4. 18K Rose Gold Double Hearts bracelet on Chain FC-505A9
5. Silver Pendant Double Hearts FC-203A1
6. Bicolor Double Hearts bracelet FC-702A19
7. Nicole Faria Miss Earth & India 2010 Brand Ambassador

1

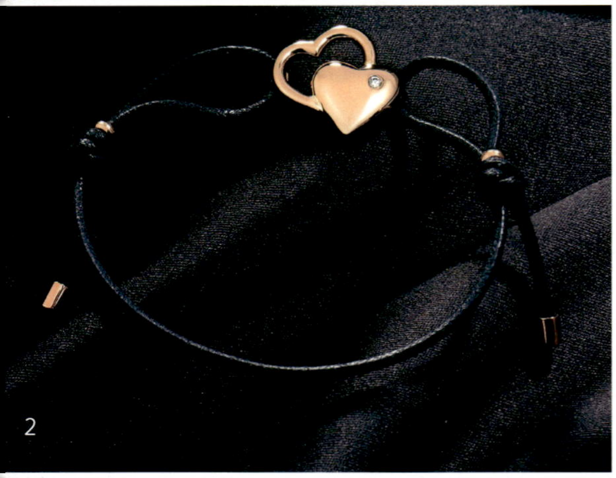

2

3

4

5

6

Women love jewelry. Frédérique Constant now also includes jewelry in its collection. How did that come about?
ALETTA: I had been aware for quite some time that watches and jewelry both do and don't belong together. Specialized shops usually sell both. A wristwatch also plays an ornamental role. But jewelry per se is another story. Peter and I were on vacation in Italy and I asked him whether we perhaps ought to consider adding jewelry to our portfolio.

PETER: We jotted down a few ideas and we talked about the possibilities of making something that would complement our "Double Heart Beat" watches. We weren't considering jewelry as an independent pillar in our product spectrum, but we wanted to make something complementary that would further strengthen the genuinely iconographic image of those ladies' watches. When you look in jewelers' display windows, you often find watches from one brand displayed alongside jewelry from another. That needn't be so. Here was where we could offer something appropriate with Frédérique Constant's name on it.

When did you first start thinking about Frédérique Constant jewelry?
ALETTA: In September 2010. We held the first pieces in our hands three or four months later.

Do you collaborate with Italian manufacturers?
ALETTA: No, our producer is in France.

How well were the new items received by your clientele?
PETER: The response was unfortunately somewhat tentative. We typically have a distributor between us and the jewelers. Many of them are men and therefore somewhat hesitant.

ALETTA: We officially launched Frédérique Constant jewelry at Baselworld 2012. The products were immediately successful in the Middle East. The first deliveries are just now coming into display windows in Switzerland, so it's too early for me to make a generalized statement about the reception. We've received some orders, but success is measured in terms of sales to retail customers. It's also important that jewelers enjoy

JEWELRY
Double Heart Jewelry collection: when two hearts beat as one...

combining our ladies' watches with jewelry that we've expressly created to complement those timepieces. We're pleased that the initial responses have been positive, so I'm looking to the future with optimism.

Do you have your own designer for these products?
PETER: Not right now.

ALETTA: But I can reveal that I already have ideas for new pieces of jewelry which will premiere on the market in 2013.

Have you set an upper limit for the prices of jewelry bearing Frédérique Constant's name? Or are you thinking, for example, about lavishly diamond-studded little treasures?
PETER: From a marketing point of view, it would be unrealistic to start too large and too high. Our jewelry's prices will always have to be commensurate with our watches' prices. As I said, right now we don't want to make jewelry into a second pillar of our product portfolio. We're presently focusing on the area of expertise in which we truly excel -- and that, of course, is wristwatches.

ALETTA: Jewelry is only one facet. There's an old saying: "Never say never." Perhaps someday we'll also make exquisite one-of-a-kind pieces. But for the present, our motto remains accessible luxurious jewelry to match our accessible luxurious watches.

What sort of items do you have for gents to give to ladies and for ladies to give themselves?
ALETTA: Bracelets, necklaces and earrings.

No rings?
ALETTA: No, definitely not. That's much too elaborate because of the various ring sizes and the large stockage which that entails. In every instance, Frédérique Constant's jewelry must fit its wearer and she must be able to take it with her immediately.

What materials do you offer?
PETER: Solid silver plated with rhodium to prevent tarnishing and 18 carat rose gold. We decided against white gold because of its high price. We want to and we must remain faithful to our philosophy of accessible luxury. Silver with diamonds is also a gorgeous combination. The price is right too. A lady can truly afford it.

What about stainless steel?
ALETTA: That material, in combination with diamonds, has enjoyed a very positive evolution during the past several years. Stainless steel is a high-quality substance and it looks good. We're keeping it in mind as an option for the future.

Frédérique Constant's strong success must make your company desirable for potential buyers.
ALETTA: Yes, indeed. About every two weeks we receive a letter from a private equity investor or a hedge fund or a consultant. But those letters are immediately thrown into the wastebasket. Two such groups approached us in the past. We thought about it for a few minutes, but ultimately it was clear that selling our business is out of the question.

PETER: We still have work ahead of us. We're growing continually and our status as a family business helps us grow. I was 48 when we were asked about selling. I thought it over, recalled my experiences at Philips and realized that I probably wouldn't function very well in a group with political structures. So we decided to continue on our present path for the foreseeable future and for as long as it remains a viable option.

A family business is optimal, but even a family-run business needs qualified managers.

PETER: Yes, we've instituted hierarchical structures. There's a production chief, a commercial director, a financial chief, a technical director, a head of communications, a product manager and a director of the R&D department. In this way, we've formed a management team.

According to which criteria do you proceed?

PETER: There's something interesting to report about that. I think it was 2007 when I joined the Young Presidents Organization (YPO). It was originally an American association, but now it has about 70 members in French-speaking Switzerland alone. They organized a seminar in Vienna with the title "Build Your Own Management Team." That seminar inspired us to implement such a team at Frédérique Constant. Afterwards, we and the entire team traveled to a YPO event entitled "Rockefeller Habits." (1)

What does that mean?

PETER: One of the methods is to hold a meeting of the executive team in my office every morning from Monday through Thursday at 8:58 a.m. Each person has exactly one minute to describe what he or she intends to do on that day. A manager only needs to mention a number and everyone knows how far along a project has come, e.g. watch production or the quantity of published articles. The third round considers obstacles of all sorts: marginal problems can be discussed and solved immediately; larger difficulties are addressed during the course of the same day, so the next morning's meeting begins with a new situation. That's not only extremely efficient: it also quickly and significantly improved the functioning and cohesiveness of our management team.

How long do such meetings usually last?

ALETTA: Fifteen or at most twenty minutes. They really go very quickly.

What do you do next?

ALETTA: Each of the chiefs goes to his team and holds a similar meeting. Here too, the entire conference takes between ten and maximally twenty minutes.

PETER: I believe it's very important for such meetings to be held every day.

Have these methods proven themselves at Frédérique Constant?

ALETTA: Absolutely. Everyone has an opportunity to speak. Colleagues get to know one another better. Each one can tell the others where the shoe is presently pinching, so to speak. There's no mystery-mongering.

Is Aletta also present at these meetings?

ALETTA: Yes, at least three days each week. I stay at home on Wednesdays, but I'm virtually present at the meeting via teleconference.

Do the two of you jointly choose your personnel?

PETER: We usually make hiring decisions together, but our business has become so large that we can no longer personally make every personnel decision. We've delegated responsibility to a specific individual in each department. I usually contribute to the decision when we're hiring a new person for our department.

You studied law, Aletta. Do you deal with the contracts for Frédérique Constant?

ALETTA: We use a standard single-page contract for hiring new employees. The wording is terse and to the point. I personally reworked the contracts between us and our agents, but to be quite honest, I don't really love that task Legal things always have a slightly negative touch which doesn't appeal to me. I prefer to move things in a positive direction, so I'm very happy that I'm no longer active in the legal profession. I find it much more appealing and more exciting to devote my energies to watches.

(1)

YPO
Logo of the Young Presidents Organization

COMPANY TEAM
Collage of photos of personnel made for the 5th anniversary of the Manufacture in Plan-les-Ouates.

Fifth Anniversary of the Manufacture

Plan-les-Ouates, Geneva - June 16th, 2011

Our team character is very strong.

From an employee's point of view, is there a significant difference between working for a family business and being employed by a big corporation?

PETER: Yes, I would say so. Our structures are leaner and our hierarchies are flatter, so anyone who's promoted to a higher position needs to think and act differently. In a certain sense, he also has to act as an entrepreneur.

ALETTA: Our colleagues must make more decisions on their own. They can't hide behind a superior's back. That's not easy. Furthermore, there's clearly a difference between reporting to a hired manager and reporting to the business's owner.

Does everyone at Frédérique Constant feel as though he or she is part of the family?

ALETTA: I can only speak for myself, not for our employees, but I certainly feel that way. Together, we're strong. We work hard, but we also have fun together now and then. We highly respect one another. Achievements are noticed, praised and rewarded. This could also be the situation in a business that isn't run by its owner, but it's especially obvious at Frédérique Constant. Our team character is very strong. We try to act in a straightforward way. We avoid political intrigues. If we discovered political wrangling, we wouldn't tolerate it.

PETER: That's not solely due to the familial aspect. Our Dutch mentality also plays a role. We Dutch tend to be open and direct. We prefer colleagues who speak openly and tell us how they feel rather than buttoning their lips and nursing a grudge. Whenever an issue arises, we discuss it and make every effort to find solutions.

How often do you meet Peter each day?

ALETTA: When I'm there, which is usually three days a week, we often go to lunch together to discuss the current situation – and there are plenty of situations in the course of a week! It's hard to believe. Many issues can be resolved in just five minutes. But the time together is important. If there isn't enough time to go out for lunch, we'll eat a sandwich in the office and talk about the business right there.

Peter is the CEO. What are your specific tasks at Frédérique Constant, Aletta?

ALETTA: Peter and I clearly separate responsibilities. That too is important. I direct the operative segment as COO.

What does that involve?

ALETTA: Everything except sales and marketing. Finances affect me only indirectly because of payments. From that viewpoint, I feel responsible for 125,000 watches each year. They must be manufactured, shipped and subsequently maintained or repaired by our customer service team. There's always plenty to do. Everything is a lot of work. But here it has to do with interaction between people and production processes, where technical problems can always crop up. That's why for the past year – and I really should have started this earlier – I've held regularly scheduled meetings with the heads of the departments. We discuss difficulties and we search for practical solutions. They often tell me, "We'll take care of it ourselves," but I believe that impulses from my point of view are very important. The construction sites grow alongside the growth in the complexity of our production, and with T1 that also includes the assembly of watch movements.

But you're neither a watchmaker nor a technician.

ALETTA: That's why it took some time before I understood the various sequences and the interaction of the numerous components. The basics are the same, but the details change constantly and the solutions must clearly relate to those changes. If you approach things without being specific, sequences can falter.

Is it more difficult or is it easier for a woman to interact with the mostly male engineers, technicians, watchmakers and craftspeople in a watch factory?

ALETTA: I don't experience it as a problem. One can collaborate outstandingly well with technicians. They're relatively uncomplicated.

ALETTA STAS
Presentation of new products
to the NY Press

FREDERIQUE CONSTANT
GENEVE

FREDERIQUE CONSTANT
GENEVE

SMART BUY

A modern luxury culture which is fueled by commercialism seems to advocate purchases of luxury items at very high prices. Some people seem to focus only on how expensive an item is, rather than focusing on its value. This unfortunate practice has eroded the true traditions and meaning of luxury, where price was ultimately the function of the intrinsic value and the quality of the item being purchased.

Frédérique Constant's concept of providing accessible luxury timepieces stays true to the tradition of providing luxury items for an owner's personal satisfaction. Frédérique Constant provides value regardless of price and not the other way around, where value is used as an excuse to drive prices of luxury items higher.

Such purchases could be termed a "smart buy," because the buyer knows he owns a respected and recognized item prized for its value rather than its cost.

As smart buyers, owners of Frédérique Constant timepieces are satisfied. They have acquired products of luxury that were developed by passionate watchmakers with a perfectionist's eye for even the smallest detail. They know the value of what they own.

CLASSICS MANUFACTURE
FC-710MC4H6 (2012)

ACCESSIBLE LUXURY

By definition, accessible luxury refers to luxury items which are available at an accessible price. But to some people, the idea of accessibility seems alien when the subject of luxury is discussed. Take, for example, the undiscerning buyer who might have the financial wherewithal to purchase an expensive product without understanding the finer details of its intrinsic qualities. This sort of person understands luxury products merely as expensive items which are largely unavailable to most people. With this in mind, why should anyone want a luxury item which is available to a greater number of people at an accessible price?

The answer to this question lies with the discerning buyer of luxury items, who understands that luxury products are not just about their prices alone, but mostly about the quality of the items purchased. Quality refers to the materials used in making the product, as well as the refinement and extent of the craftsmanship involved and the reputation of the brand behind the product.

With items such as luxury watches, it is widely known that the best items in this niche are largely made in Geneva, Switzerland, a city renowned for watchmaking practices and innovations which have for centuries held the attention of those who understand the nature of timepieces.However, some brands have taken advantage of this reputation to develop expensive products on a purely commercial basis, which limits the purchase of such items to a very few, including those buyers who view cost as being the sole criteria for identifying items of luxury.

These practices are usually justified as being due to the rarity of the watches in question, the number of items made for a particular collection, as well as the cost of the precious materials used in their manufacture.

But a watch buyer should always be able to relate the cost of a product to the value of the item purchased. Our focus at Frédérique Constant is on creating luxury timepieces with an intrinsic value that's usually greater than the price of the products being sold. Intrinsic value involves the quality of a product in terms of its design, the materials used in its manufacture and strict compliance with Swiss watchmaking standards and traditions.

Our timepieces are exclusively manufactured by hand. The processes involved are complemented with the latest high-precision equipment available in the industry, backed by years of research and development. We recruit passionate watchmakers from time-honored watchmaking schools whose reputations are an integral part of Swiss watchmaking history.

Unlike a number of other brands worn for name recognition and price tag alone, you will rarely find a Frédérique Constant watch on the wrist of an individual who is not also a connoisseur of the finer points of watches and life in general.

Frédérique Constant offers watches as accessible luxury items, while maintaining a perfect and delicate balance in terms of pricing, quality and availability. Making watches with beautiful designs and high quality is our passion. Lovers of our brand and what it stands for are partners in this shared passion for fine timepieces.

CLASSICS
Chopin Limited Edition, tribute to the famous pianist Frédéric Chopin

OUR PURPOSE
LET MORE PEOPLE ENJOY LUXURY

Let us explain in greater detail what we mean by Let more people enjoy luxury.

More People
We do not want to restrict interest in Frédérique Constant watches to a limited and elitist circle of connoisseurs, but rather we want to open it to a broader selection of appreciative enthusiasts who want to enjoy Swiss high quality classical watches at sensible prices. Frédérique Constant accordingly strives to democratize luxury.

Enjoy
We aim to share our passion for fine watches and let people enjoy the pleasure of owning and experiencing beautiful and well-finished products. Our customers appreciate the quality and refinement of our watches, which make them feel good and happy. Frédérique Constant watches create memories of special moments.

Luxury
A Frédérique Constant watch has all the attributes of a luxury timepiece, but sells for an accessible price. Frédérique Constant watches are defined by their high quality, differentiation and precision in design and manufacture. We have a passion for designing our beautiful watches. Their perceived value — through quality of design, materials, and manufacture — is a key component of our success. Each of our watches is assembled by hand and extensively controlled with the latest equipment to ensure maximum quality. We innovate to offer creativity and exceptional value.

FRÉDÉRIQUE CONSTANT'S CORE VALUES

ACHIEVEMENT

At Frédérique Constant, we believe that for any company to be truly successful, it must aspire to, establish and abide by a strong and dynamic set of core values.

We see our core values not as rigid guidelines that limit or inhibit our workforce, but as a creed that constantly encourages our dedicated and talented employees, and reassures our customers that Frédérique Constant will never be satisfied merely with success. We aim for more... much more... which is why the first of our clearly defined core values is "Achievement."

How do you measure achievement?

Achievement can be measured in many ways. At Frédérique Constant, we look at all these standards of measurement and then we look for more.

Growth drives our company, because through growth, we can continue to encourage and invest in the many things that are important to Frédérique Constant. Over the past quarter of a century, our company has experienced unrivalled growth in the Swiss watchmaking industry, averaging a 25 percent increase in sales each year during that time.

The incredibly talented artisans who design and assemble Frédérique Constant watches have won numerous awards. They come to work every day with a winner's mentality, knowing that they are important parts of a highly successful company.

Our employees' achievements have earned recognition and rewards. Because of their success, the company has been able to invest in an ultra-modern, 32,000-square-foot facility with an optimally attractive, dust-free and climate-controlled working environment.

The growth in sales of Frédérique Constant watches has enabled our company to invest in the newest, most technologically advanced equipment in the industry, helping to ensure and sustain future growth and job security for our employees. Achievement has led to success at Frédérique Constant, and this success is one of our greatest achievements.

SHUQI
Brand Ambassador ShuQi in Beijing Children's Hospital

ICHF
Dr William Novick
Cheque for the International Children's Heart Foundation

(1)

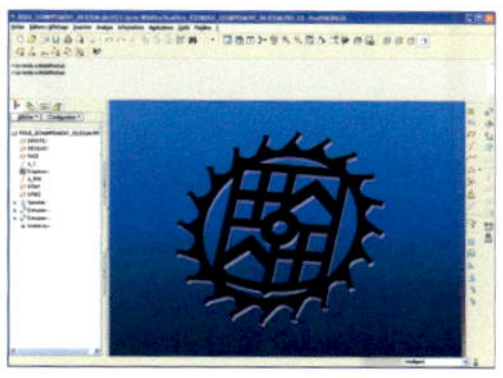

SILICIUM
Proprietary Silicium escapement wheel developed on Pro/ENGINEER software

WITH SUCCESS COMES RESPONSIBILITY

Rather than rest on our laurels and concentrate solely on sales and profits, the men and women who are the driving forces at Frédérique Constant have always believed that successful companies have a responsibility to give something back to people at home and abroad, and to the planet on which we live.

We understand that we have been fortunate and we see it as our duty to do our utmost to help those who are less fortunate than we. As such, Frédérique Constant has been a passionate advocate of and contributor to a wide variety of charities, particularly those that deal with children, education and heart disease.

To focus attention on ecological issues, Frédérique Constant has joined forces for the first time with "Miss Earth," Nicole Faria of India. Miss Faria is extraordinarily passionate about current issues, such as global warming, pollution and environmental hazards, which affect the entire planet. When she heard that Frédérique Constant is a keen supporter of ecological initiatives, she was quite happy to collaborate with us to raise awareness even further.

In China, the company joined forces with one of the country's greatest movie stars, Miss ShuQi, to visit young patients in the cardiology department of Beijing Children's Hospital. Frédérique Constant also sponsored the redecoration of the hospital's wards in conjunction with the "Paint a Smile" Foundation.

ACHIEVEMENTS IN INNOVATION

One of the greatest achievements to which any watchmaking company can aspire, is to create and produce its own calibers. In the past eleven years, Frédérique Constant has managed this achievement with not one, but four calibers in its Manufacture Series of watches. Our company is one of the very few Swiss watch manufacturers that can actually claim to produce and use its own calibers in-house, a fact which is even more amazing considering that Frédérique Constant is a completely individual, family-owned business, not part of some giant, faceless corporation.

We also pioneered a new development when we produced a silicon escapement wheel for use in the Manufacture Tourbillon series. Stronger than steel and never requiring lubrication, the silicium escapement also provides a degree of accurate timekeeping previously unattainable in mechanical watches. (1)

PASSION DRIVES US TO ACHIEVE

"Live Your Passion" is the company's slogan, but it doesn't apply only to our customers. Every watchmaker who works at Frédérique Constant is truly passionate about what he or she does for a living. This is what we encourage and what we expect from each of our employees..

Passion is what makes the heart beat faster, so what better way to demonstrate our passion than through our most successful watch ever? In 1994, we developed our first Heart Beat watch. (2)

We wanted our customers to be able to see the inner workings -- the heart and soul -- of our mechanical watches, so our designers placed an aperture directly over the balance wheel. This allows the proud owner of a Heart Beat watch to see the intricate movements under the dial.

Over time the aperture was modified into the shape of a heart. We think it's the perfect symbol of the passion that goes into the design and assembly of every Frédérique Constant watch.

GREATER ACHIEVEMENTS TO COME

Our goal is not merely to achieve, but to continue to achieve, to succeed and to win for generations to come. Only through continued investments in our workforce and new technologies can we be assured of future success.

(2)

Damalige Broschüre (1994)
Earlier brochure (1994)

By continuing to achieve, we can also continue to do what we're most passionate about: helping others and giving back to the global community. Because this is our greatest achievement of all.

ALEXANDRA SANCHEZ
Customer Service Manager

INTERVIEW WITH ALEXANDRA SANCHEZ ON "ACHIEVEMENT"

Alexandra, were you already familiar with the Frédérique Constant and Alpina watch brands before you joined the company?
ALEXANDRA: I had heard of them and I had seen a few photos or advertisements.

How did you come to Frédérique Constant and Alpina?
ALEXANDRA: I had worked in Spain, at first for Club Med, and then for a freight forwarder. A girlfriend in Geneva told me about the world of Frédérique Constant and recommended me to the company. The time was ripe in 2009, when a new phase in my life began.

You now play a truly central role in this company.
ALEXANDRA: Yes, you certainly could say that. All incoming queries from customers around the world cross my desk. As manager of Customer Care Worldwide, I communicate with the whole world every day. I'm the contact for every problem. I'm also responsible for dispatching the spare parts that are needed for repairing our watches.

How many languages do you speak?
ALEXANDRA: Four: Spanish, French, English and Italian.

Are those four enough?
ALEXANDRA: Sometimes I wish I could speak a few others.

With Spanish, French, English and Italian, how do you handle worldwide communication? For example, how do you communicate with Russians?
ALEXANDRA: I speak French with our Russian customers and English with our Chinese clients. Communication with India can be a bit difficult because not all Indians speak the Queen's English.

When you start your work each morning, some of your customers are finishing their workdays and others are still asleep. I imagine it isn't easy to keep everything synchronized?
ALEXANDRA: Yes, some of my workdays can be very long. But experience has shown that queries from around the world can be managed from here in Geneva.

Sometimes I'm more successful, sometimes not at all. It can be frustrating, but I'm always learning, especially when I've had a trying day.

You work at a very demanding job. Where do you get your daily dose of motivation?

ALEXANDRA: I don't feel like an employee: I feel like an entrepreneur within this entreprise. And I really like to work. Peter and Aletta Stas give us all plenty of latitude to do our work as we see fit. What ultimately matters is that everyone does their job and contributes to the firm's success.

Yes, Aletta and Peter Stas place great importance on independent work and on personalities in their company.

ALEXANDRA: That's probably one of the secrets of our success. I take responsibility for positive communication with the outside world. I must be able to rely on a perfectly motivated team in the background. Big challenges constantly strengthen our solidarity. Day in and day out, we must remain vigilant so that our spirit and the company's culture always remain in harmony. All of us here live our principles. We respect one another. But sometimes we can feel a bit overwhelmed. The good thing is that at the end of the day, it's tremendously motivating to perform necessary tasks at or above one's limit.

Alexandra, you still work alone, that is, you have no assistant to lighten your workload. Will things stay that way?

ALEXANDRA: As I said, working at the limits of my capacity is highly motivating. It makes me very happy to look back and see what I've accomplished. My capacity isn't the problem. But we're growing so quickly and my sector has grown accordingly, so one individual alone can no longer handle it all on her own.

You've therefore posted an ad on the internet to seek an assistant who'll reduce your workload. Was that easy, considering how people work at Frédérique Constant and Alpina?

ALEXANDRA: No! Let's take a look at that ad together. I think it couldn't have been phrased more clearly.

Yes, the text leaves no doubt about what sort of person the company is looking for. The first item in the "preconditions" category is that the future Customer Service Assistant must be able to work in a highly proactive manner and should not come to the job with a nine-to-five mentality. Hard, disciplined, independent work is the credo more than once. Are you basically looking for another entrepreneur to join the enterprise?

ALEXANDRA: Exactly. He or she should be someone who fits into our company's culture. More than 2,500 individuals applied to fill the post of Customer Service Assistant. Only fifteen made it onto our short list. That alone makes it clear that many candidates weren't willing to work on our level. Furthermore, taking responsibility and working independently isn't everyone's cup of tea.

What do you do to relax?

ALEXANDRA: I love nature, so I spend every free minute outdoors. I walk and hike in the great outdoors. That's how I fill my tank with fresh energy.

Frédérique Constant's Core Values

HONESTY

At Frédérique Constant, we believe that to be truly successful, both in business and in life, we must also always be open and honest: with our customers, our agents and distributors and, just as importantly, with our dedicated and talented employees. "Honesty" is therefore one of the core values which we as a company hold dearest.

IT STARTS AND ENDS WITH CLEAR LINES OF COMMUNICATION

All employees at Frédérique Constant should be passionate about what they do. This passion at all levels within the business results in a wide range of different ideas coming into the building every day. People will not always agree on what is best for the company, but every employee here at Frédérique Constant will always be encouraged to share their views, opinions and ideas at every level.

Our philosophy is that whenever anyone feels passionately about anything involving the company, they should air their point of view. Our creative designers, our talented team of artisans who assemble the highest quality Swiss timepieces, our marketing department -- these are all composed of people who bring their passion to work every day. At Frédérique Constant, we believe in listening to and acting upon the ideas and innovations that spring from these passions. We pledge that anyone and everyone within the company who wishes to share their ideas will be heard and their ideas will be considered. Passion is to be encouraged and nurtured, and this is most easily achieved through open and honest lines of communication.

FAIR PLAY FOR ALL

Every Frédérique Constant employee can be assured of fair treatment in the workplace. Politics, whether internal or external, will never be allowed to cast a pall of negative influences on our company.

The mere fact that someone is working for Frédérique Constant means that they are a highly skilled individual who displays a passion for what they do. Our employees work for a successful company which aspires toward continuing achievements within the industry in general and which honestly strives to help those less fortunate than ourselves. As such, each and every one of our employees is a valued part of our business. We give an earnest commitment to treat them honestly and fairly, and to listen attentively to their ideas.

In return, the staff and management at Frédérique Constant are always mindful to show the utmost honesty and respect to fellow employees, to our customers and to our visitors.

HONESTY LEADS TO TRUST

Our company runs on teamwork. Each individual in the company plays an important role within the Frédérique Constant team. For any team to perform as efficiently and successfully as possible, all individuals involved must trust those around them.

We strongly believe that honesty and trust go hand in hand. And we strive to provide the honesty that inspires trust in our friends, colleagues, suppliers, distributors and customers.

PUBLICITY
iPad App Frédérique Constant

(1)

Guarantee Certificates Alpina and Frédérique Constant

(2)

Frédérique Constant website

HONESTY EXTENDED TO OUR CUSTOMERS AROUND THE WORLD

We understand that without our customers, there would be no company, no matter how wonderful our watches might be. As such, Frédérique Constant is committed to treating our customers honestly and fairly, and we have well-established policies and procedures in place to ensure that this will always be the case. .

We offer one of the most comprehensive product guarantees in the watchmaking industry. Every watch we produce is covered against any defects in materials or workmanship, no matter how small, for a full two years. Any defective component is guaranteed to be replaced free of charge. One of our brands, Alpina, was the first to introduce this international guarantee. (1)

We also show honesty in our quality-control procedures. Every Frédérique Constant watch is thoroughly tested and inspected by highly experienced craftsmen and the most technologically advanced equipment available. After all, when a watch will pulse more than a billion times over four years, it's important that it be very well made!

HONESTY WITHOUT FEAR OF MISUNDERSTANDING

Frédérique Constant is a global brand and we believe that people who wish to learn more about our company should be able to do so as easily as possible. To ensure this, our website can be accessed in no fewer than eleven different languages, including both simplified and traditional Chinese. (2)

To keep abreast of changing times, styles and popular trends, the company has also created and launched its own iPad app, which can be downloaded directly from our website.

BENOIT FAVRAT
Financial Director

HONESTY SHOWN THROUGH INTEGRITY

Frédérique Constant is genuinely concerned about giving back to the community, not just locally, but globally. As such, we have been involved in a wide range of events, foundations and activities that aim to help less fortunate people, particularly children.

Just one example of this is our recent collaboration with the actress Eva Longoria and the Par Coeur Association, a French foundation which works on behalf of families around the globe. A gala was held to help fund "Toutes à l'Ecoles," a nonprofit organization that creates schools and educational programs for young girls in developing countries. *(1)*

We also donate $50 from every one of our Heart Beat watches sold to help with the research and prevention of heart disease.

"Honesty" is not just a word at Frédérique Constant -- it is a true value that partly defines who we are and how we treat the people who work with us, as well as our agents, our customers and the people who make up the world in which we all live.

INTERVIEW WITH BENOIT FAVRAT ON "HONESTY"

Benoit, as head of accounts at Frédérique Constant and Alpina, you handle all of the company's pecuniary aspects. Part of your work also means closely scrutinizing your colleagues. Are people afraid of you?
BENOIT: No. My responsibility is to conduct honest, systematic analyses that make all of the firm's financial transactions transparent so everyone knows where he stands. Preparing statistics for Aletta and Peter Stas is particularly important. I meet with our operations director Olivier Zaugg every Monday to discuss the financial situations of our firm, our subsidiaries and our partners.

You're so close to the numbers. Do you sometimes hear, as it were, the grass growing?
BENOIT: Yes. And all too clearly sometimes! It can be hard to believe. Nearly nothing can be kept hidden given the constant contact with all our subsidiaries and independent distributors. I receive reports about sales and performance worldwide. The whole system is like a finely tuned seismological network. It's sensitive to even the slightest tremors.

(1)

Toutes à l'école

EVA LONGORIA
Par Coeur Gala 2010 – Paris

Did you see the last big crisis coming before it hit in 2008?
BENOIT: Yes! We knew that things were already slipping downhill long before the media reported it. We could only guess how steeply and how fast, but we drew the conclusions soon enough.

You're also responsible for the firm's insurance and electronic data processing.
BENOIT: Those tasks round out my field of responsibilities. Peter Stas encourages entrepreneurship in our company. Each of us has greater or lesser responsibility for a given area and is expected to live that responsibility. I wear three hats and each one fits me quite well. Insurance is closely affiliated with finances, so it makes good sense for me to handle both.

Intelligently organized data processing makes many internal sequences easier. How important is EDP in a company that prioritizes togetherness and day-to-day communication?
BENOIT: Sequences at Frédérique Constant and Alpina run so smoothly because we talk so much with one another. Each of us is part of one big community. We help each other and we collaborate closely because even the most intelligently organized data processing is useless without these fundamental values. But when those virtues are practiced, then the computers are as powerful as jet engines.

How did you find your way to this company?
BENOIT: I still shake my head in disbelief every time I tell this story. Time and again, I ask myself: "Did you really quit working for Rolex?" Yes, that's exactly what I did! After my financial education, I had a temporary job at Rolex. I was supposed to work there for four months. Afterwards, my performance would be evaluated and they would decide what my future would be. The fact of the matter is: if you don't make any stupid errors during those first four months, then chances are good that they'll offer you long-term employment. But I'd had enough of Rolex after just two weeks. Everything there is hierarchically organized. The basic idea is that no individual should bear any responsibility. That ran against my grain because individual responsibility is exactly what I wanted. I yearned to be able to work independently and to think for myself. My mother recommended that I should consider Frédérique Constant, and I was hired soon afterwards.

When was that?
BENOIT: Ten years ago.

That was indeed a courageous step because no one at that time could have predicted what Frédérique Constant would ultimately become.
BENOIT: I liked – and I still like – the candid, straightforward way we work here together with Aletta and Peter Stas. My warmhearted, open and honest colleagues contribute their share too.

What do you do now and then to let off steam?
BENOIT: I jog regularly and eagerly. I live in the city of Geneva, which is very inviting for joggers. It's wonderful to relax beside the lake.

Frédérique Constant's Core Values

EFFICIENCY

As a company, we have a great many ambitions. We are driven to succeed in business on a global level and we are determined to remain a leading representative of the Swiss watchmaking industry for generations to come.

We also have a strong desire to make a positive and lasting impact on the global community at large. To this end, we are constantly looking for new social and charitable projects with which we can become intimately involved and particularly those that help people less fortunate than ourselves, both locally and abroad.

To achieve all this and more, and to remain profitable while doing so, Frédérique Constant recognizes the need for the company to show the highest levels of "efficiency" at all times and at all levels within the business.

TIME WAITS FOR NO MAN... NOT EVEN FOR THE WATCHMAKER!

Frédérique Constant produced and sold more than 110,000 watches in 2011 alone. We strive for growth every year, but as a company we make a commitment to our customers and agents that we will never achieve growth by cutting corners or by taking shortcuts in production.

Our quality assurance procedures and standards will never be cut back. As our company grows, we will facilitate increases in production with higher levels of efficiency, but never at the expense of the unrivalled quality of the watches we make. We know we cannot stand still, because there remains much for us to achieve as a company. Time marches on. For Frédérique Constant to keep up, we will have to become even more efficient. (1)

Our company's growth must be sustainable. And achieving consistent improvements in the manufacturing and sales of our high-quality timepieces will require every member of our staff to remain focused on the company's goals and how they can help us to achieve the success that ultimately benefits everyone connected with Frédérique Constant.

Our employees are known for the passion they bring to work each and every day. This above all has driven Frédérique Constant forward and into a position where the company is one of the leading Swiss watchmakers. We remain focused on the job at hand and on our responsibilities to our employees, our agents and distributors, and, of course, our valued customers. Through even greater efficiency in every aspect of our business, Frédérique Constant can maintain its market-leading status and continue to look after those things which are most important to us, both as a company and as individuals.

(1)

CYCLOTEST
Quality Control machine to check reliability of each watch for 48 hours

(1)

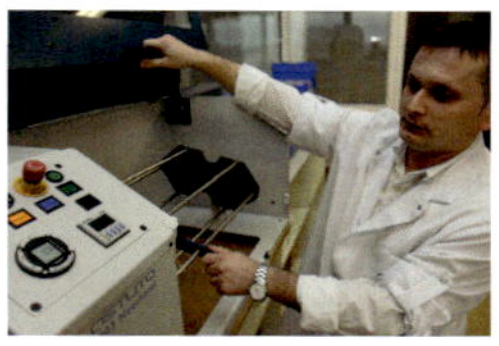

FAST TUMBLER
Quality Control machine to check integrity for 5 hours

FAST AND FLEXIBLE

For Frédérique Constant to continue to grow as a company, we will need to show flexibility and the ability to adapt to new production techniques when required. We know that we can become faster and stronger as a company without sacrificing the quality that is crucial to our success.

We encourage our skilled and passionate workforce to voice any ideas they may have on how the company can improve in any area, and the management team at Frédérique Constant is happy to give a commitment that we will listen to and respond to those ideas. Where appropriate, we will also test and implement new strategies that can improve the speed of production processes and reduce costs without sacrificing quality at any stage.

EFFICIENT TESTING TO ENSURE CONTINUED EXCELLENCE

Frédérique Constant watches are not just beautiful pieces of jewelry. They are also incredibly accurate and durable timepieces, and we take extensive steps to ensure that each and every one of the more than 110,000 timepieces we manufacture each year upholds the highest technological standards.

To do this, we submit every watch we make to strenuous quality-control examinations, including placing each watch in a fast tumbler for five hours to simulate the shocks and vibrations it will encounter in its lifetime. We also put each timepiece through a 48-hour slow rotation test to simulate normal wrist rotation, as well as a water resistance check, visual inspection by trained craftsmen and a time-accuracy test. (1)

Frédérique Constant uses the most technologically advanced equipment in the industry and the most efficient testing methods to make sure every customer is delighted with the finished product.

PASSIONATE CAN STILL BE EFFICIENT

Business requires a certain degree of practicality, particularly in the light of current global economic difficulties. However, Frédérique Constant employees at every level should never assume that being practical and efficient means they must also dampen their passion and enthusiasm. We feel that quite the opposite is true. By adapting to current trends and becoming more efficient, we are free to focus our passions even more sharply on the real reason we are with this company: namely to create the finest watches in the world.

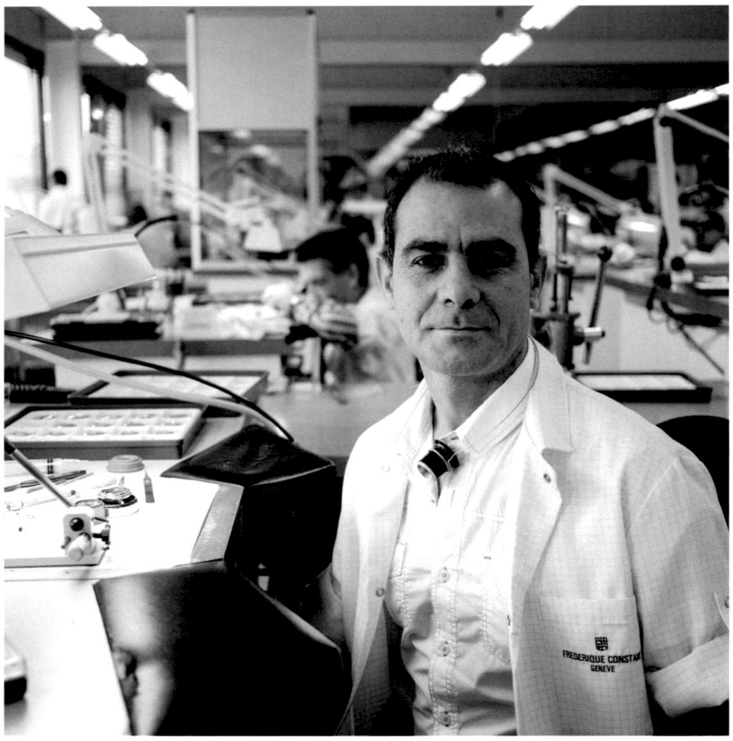

JOSE DOS SANTOS
Watchmaker in T2 Atelier

INTERVIEW WITH JOSÉ DOS SANTOS ON EFFICIENCY

You began encasing and assembling movements at Frédérique Constant ten years ago. Before that, you were a baker. Was the change from one profession to the other difficult?

JOSÉ: I'm 43 years old. This is only my second job. Before that, I was a baker for 14 years. The plan was for me to undergo retraining, but I was literally thrown into the midst of things. Practice was my teacher. The good part was that I was very interested in my new task, so I paid strict attention, asked the right questions, understood quickly and progressed rapidly.

Had you had anything to do with watches before you joined Frédérique Constant?

JOSÉ: No, my only point of reference was my Swatch chronograph. Before that, I'd never even changed the battery in a watch. When I worked as a baker, I didn't even wear a wristwatch.

José, the way you work is uncommonly efficient. Scarcely anyone at Frédérique Constant can assemble as many ladies' and jewelry watches per day as you do.

JOSÉ: At first, I was able to put together no more than twenty watches each day. The total gradually increased and I can now make an average of fifty per day. If everything goes smoothly, I can sometimes assemble even more.

What motivates you to work so fast?

JOSÉ: I enjoy my work. It gives me pleasure every day. If you work with good concentration, it's not a problem. You don't waste valuable time. For us in the assembly department, this means we must rigorously comply with our guidelines. Efficient work is the driving force! One colleague may complete more watches per day and another may make a few less. That's life. But we're all equally industrious. There's also something else that we share in common because we work for a company

that manufactures many thousands of watches each year. Time waits for no man, not even for a watchmaker.

That sounds a bit like constant competition among the members of your team.

JOSÉ: No, we don't compete with one another. We all try to work efficiently and cope with the daily pressure as best we can. It functions quite well, also because 12 of my 16 colleagues in atelier "T2" have been here for many years. There are few fluctuations on our staff.

That sounds like an ideal working atmosphere.

JOSÉ: That's exactly what it is! We're a team. We stick together. We also get along very well with one another in our private lives. A positive atmosphere in the workplace motivates us and encourages solidarity in good times and in bad times. Thus each of us can work efficiently and to the full extent of his or her ability.

How often do you have contact with Aletta and Peter Stas?

JOSÉ: Aletta and Peter are frequently present. We see them often. Each of us can come to them with an issue or a suggestion whenever we have one. Especially when our suggestions involve improvements in the sequence of the work, our ideas are thoroughly discussed in the group. Communication at Frédérique Constant and Alpina isn't a one-way street and we don't feel like employees in a big corporation. We feel much more like integral parts of a whole. Each of us bears responsibility for what we do.

Do you sometimes feel nostalgic and miss the baker's métier?

JOSÉ: No! I like to bake bread for my family. I'm a family man, and I like to make my loved ones happy with my baking.

That brings me to my final question. What do you do when you're not working?

JOSÉ: As I just said, my family is my biggest hobby. We recently moved into a house with a garden and that keeps me busy. To relax, I like to play sports and to do little craftsmanly jobs around the house. When you own your home, you can always find something to do!

Frédérique Constant's Core Values

CREATIVITY

Producing precision timepieces like those made at Frédérique Constant requires exceptional skill and, of course, the finest materials and most advanced techniques. Our company is determined to maintain the highest possible production standards in every watch that bears the Frédérique Constant name.

We understand the importance of producing reliable, durable and functional watches that are as technologically advanced as they are well-built. Behind the engineering skills and quality materials, however, is a core value which we hold especially dear at Frédérique Constant, for without it, we would merely be repeating what we have done before, or worse, we would end up imitating others.

That core value, which goes hand in hand with the passion that truly defines our company, is "Creativity." We feel it is not enough that our watches perform brilliantly, with unrivalled accuracy, and that they are durable enough to be passed from one generation to the next. We also want our watches to be beautiful, even inspirational works that reveal to anyone who sees them the creative genius that led to their development as a Frédérique Constant product.

CREATIVITY THAT LEADS TO GREATER BEAUTY

Frédérique Constant employs exceptionally talented designers to produce ever-changing models that are contemporary and modern, that reflect changes in fashions and popular trends, yet are, at the same time classical and timeless. A Frédérique Constant watch will never look outdated or unfashionable because true beauty never ages.

We also occasionally draw upon the talents of independent designers who are only too happy to help us create a specifically themed model. As a company, we don't allow ourselves to believe that Frédérique Constant has a monopoly on creativity, we will seek it out from the finest sources available.

We use the most modern Computer Assisted Design software available in the industry, but we know that any such program can only be as creative as the mind of the human who works with it. We truly believe that we have the most passionate, creative minds in the business working here at Frédérique Constant. (1)

(1)

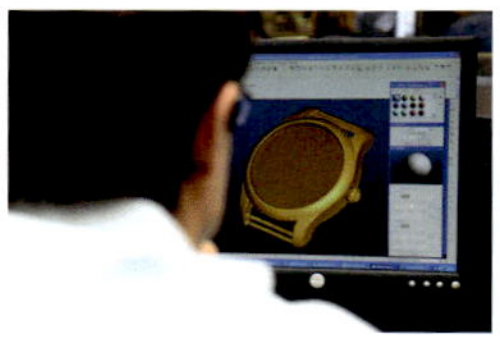

MANUEL DA SILVA MATOS
Development of a watch case on Pro/ENGINEER software

(1)

FIRST MANUFACTURE FC-910MC4H9 (2004)
Cage with 88 components, in total weighing less than 3 grams

(2)

ESCAPEMENT WHEEL
Innovative Solution...
Silicium escapement wheel

CREATIVITY TO MAKE US BETTER

We never take our position in the Swiss watch market for granted. Many companies would like to occupy our place at the head of an elite field, and our designers and manufacturers make incredible efforts to consolidate what we have worked so hard to achieve.

One example of how creativity helps us retain our valued customers, as well as attract new Frédérique Constant clients, is the much higher level of detail and refinement we add to our watch dials than other Swiss watches in the same price range. Our rounded cases are also hand polished at Frédérique Constant, showing not just creativity but also our constant attention to getting the details right.

CREATING INNOVATIVE SOLUTIONS

Creativity isn't limited to simply designing watches that are more beautiful. At Frédérique Constant, we are constantly seeking new and innovative ways to make our timepieces better, inside and out.

For more than ten years, our company has produced our own calibers in-house, and Frédérique Constant has patented no fewer than five new caliber inventions. A prime example of this kind of creativity can be found in the silicium escape-wheel, which we developed in 2007. Silicium is not only considerably more robust than any material used before: it also gives mechanical watches a degree of accuracy never thought possible. (1)

In 2008, we went one better by introducing a tourbillon watch with a silicium escape-wheel -- the first of its kind ever made. The independence of producing and using our own calibers sets Frédérique Constant apart from other small, family-owned companies. This independence also frees us to become even more creative because we are not constrained by large corporate bureaucracy. (2)

IMPROVEMENT THROUGH CREATIVITY

As well as creating new designs and innovative feats of engineering, everyone connected with Frédérique Constant is encouraged to draw on their own creative inspirations to improve what might already seem exceptional to some. Our Heart Beat range for ladies is a classic example.

The Ladies Heart Beat was already a hugely successful and award-winning member of our range, but we were creative enough to make it even better. While having coffee at home one day, we were discussing how to make the appearance more feminine and more appealing. At the time, the watch featured mother-of-pearl dials with diamonds and a circular opening at "12 o'clock." Aletta suddenly asked Peter if, instead of the round opening, we should consider using heart shapes. Without even waiting to get back to the office, we drew the first-ever Double Heart Beat dial on our home computer.

CREATIVITY, INSPIRATION, PASSION: THESE ARE AT THE HEART OF EVERY FRÉDÉRIQUE CONSTANT WATCH

Our core values define us as people and as a company. At Frédérique Constant, we truly believe we have the most passionate, dedicated and talented minds in the watchmaking industry. Passionate hearts and creative minds inspire us to ever-greater achievements and to both personal and professional growth.

Because of our success, we find ourselves in the fortunate position of being able to help people around the world who are less fortunate than ourselves. Even in these areas, Frédérique Constant strives to be creative: to give assistance where it is truly needed in unique and dynamic ways. We give a commitment that we will continue to strive for even greater and ongoing creativity in our designs, our groundbreaking technical innovations and in the ways we do good for worthwhile causes and for people who need our help the most.

MARIE ANTHONIOZ
Graphic Designer

INTERVIEW WITH MARIE ANTHONIOZ ON "CREATIVITY"

You joined Frédérique Constant and Alpina in 2009. What were things like then? What lured you here?
MARIE: I had just earned my diploma – a Brevet Technique Supérieur – from the École de Communication visuel in Lyon when Yasmina Pedrini, our head of communications, asked me if I would like to work for her. My primary task, she said, would be to be creative and to develop attractive layouts and concepts for the company. She also told me that I could forget all about counting my working hours because it would be nothing like a nine-to-five job, but it would offer me practically every opportunity to develop myself and to work very independently.

You've been here more than three years now, Marie. Did things develop for you the way Yasmina described them?
MARIE: Yes, my work gives me a lot of pleasure. I can use what I've learned, express myself and really be creative. It certainly isn't a nine-to-five job! It's creative work in a very dynamic environment. Each of us feels like a small businessperson:. We take responsibility for what we do and we coordinate our actions to achieve our common goals.

What must you bear in mind when you create a staging for the world of Frédérique Constant and Alpina?.
MARIE: The two brands are young, trendy and positioned very differently from their competitors. We're about passion: a passionate combination of watchmaking in the traditional sense and impetuous youthfulness. We put those two together to create lasting values without an expiration date. Furthermore, our watches are made with very high-quality craftsmanship and they really look great. Sometimes I ask myself how they can possibly be so accessible. I've made a habit of writing down the prices so when I'm working on a project I can always keep an eye on which price segment we're really at home in.

It's astonishing and deserves explicit mention: Frédérique Constant and Alpina manufacture products for the entry-level and midrange price levels. Doesn't that turn your creative work into a permanent tightrope walk?
MARIE: Yes, it's difficult to tiptoe along the narrow ridge between accessible luxury and luxury. Our communi-

(1)

Advertising Visuals (2012)

cation and every layout must take that into account. But our message is designed so that it'll be understood worldwide. Beautiful cars and boats, for example, elicit emotions everywhere. The idea is to paint a picture of the world in which one wants to find oneself. We mustn't be too complicated. Slogans like "Live your passion!" for Frédérique Constant or "Since 1883" for Alpina are exactly right. (1)

You also mentioned other universes – cars, boats, and recently also airplanes – which likewise contribute their fair share.

MARIE: That's absolute right. But giving is also an integral part of our position. Our customers give generously to us and we return part of those earnings by giving generously to charity. It's more than a duty: it reflects our fundamental attitude in the world of luxury.

Frédérique Constant recently debuted a very well-made iPad app. Was it your brainchild?

MARIE: It resulted from exciting teamwork with Yasmina! Naturally I contributed a large part. Our goal was to adapt the iPhone app for the iPad so that it could fully exploit the potentials of a tablet computer. I tried to imagine myself as the user. How would he navigate? What would he want to experience on his iPad? Those were only a few of the questions that I asked myself. I was finished just one week later and the programmers could begin their work.

The iPad app clearly reflects your creative process. The user experiences and sees that this app was developed with creativity and common sense.

MARIE: That's exactly what our job is all about. If there isn't enough creativity, messages become artificially complicated for lack of a better alternative.

May I ask how you relax and recharge your batteries when you're not working?

MARIE: Of course. My sister and I share a horse. She lives for equestrian sports and I ride to relax. We share wonderful hours and we have a lot fun together.

Frédérique Constant's Core Values

COOPERATION

For any relationship in life to be truly successful, there must be a certain amount of give and take. A good relationship is a two-way street, whether the relationship is personal, professional or business-to-business. Any time the traffic is all one-way, or all give and no take, the relationship is bound to suffer and ultimately fail.

With this in mind, Frédérique Constant has made "Cooperation" one of the core values upon which the company commits to base all its relationships. We firmly believe that if our employees are happy in their work, if our agents and the 2,700 outlets around the world who sell Frédérique Constant watches are happy with our products, global guarantee and backup service, and if our customers are pleased with the high quality, handmade watches we produce, then the company will thrive and continue to grow.

COOPERATION STARTS IN-HOUSE

Frédérique Constant employees are more than just the talented individuals who design, build, test and market our fine Swiss watches. They are also our ambassadors, and as a company we value the passion, dedication and craftsmanship they bring to work every day. Our employees cooperate with each other in a spirit of teamwork, and by doing so they bring honor and success to Frédérique Constant as a company and to themselves as individuals.

The company makes every effort to provide a happy, healthy environment for our skilled workforce. In 2006, the same year Frédérique Constant launched the Automatic Heart Beat Manufacture range, we also opened our state-of-the-art facility in Plan-les-Ouates, Geneva.

Our skilled craftsmen and designers work in an attractive, healthy environment, with the finest raw materials and most technologically advanced equipment available in the watchmaking industry. This cooperation between company and workforce makes for a happy, vibrant atmosphere in which new ideas, passions and enthusiasm for being part of a successful company can flourish.

COOPERATION WITH OUR SUPPLIERS

At Frédérique Constant, we believe in cultivating good working relationships with our suppliers and subcontractors. We don't limit ourselves to a select few, but always attempt to find the best possible sources of the finest-quality materials and components that go into Frédérique Constant watches. As such, we deal with more than 60 individuals and manufacturing companies to source the many different parts that go into our timepieces, including dials, cases, crowns, hands, screws, rubies, springs and watch straps. (1)

(1)

SELLITA

Logo SELLITA, an important supplier movements

(1)

MANUFACTURE
FC-910AS3H9
Heart Beat, mechanical manual winding

(2)

Logos of various charitable foundations Frédérique Constant has supported

COOPERATION WITH EDUCATIONAL INSTITUTIONS

The most important component in the production of a fine Swiss watch is the skill of the watchmaker. Frédérique Constant has established an exceptional spirit of cooperation with the world's finest watchmaking schools, which happen to be located in Genève and La Chaux-de-Fonds. Many of the outstanding artisans employed by our company to produce the finest Swiss watches come from these schools. Still others learned their skills in the ateliers of other highly-renowned famous Swiss brands before joining us at Frédérique Constant.

The development of one of the most successful watches in the Frédérique Constant range took more than three years, and in the end, it was a cooperative effort that ultimately led to the desired outcome. Starting in 2001, the company began work on the manual version of the Heart Beat Manufacture, a watch that would use calibers designed and produced in-house. (1)

We worked side by side with the Ecole d'Horlogerie (Geneva Watchmaking School), the Ecole d'Ingenieurs de Genève (Engineering School of Geneva), and the Horloge Vakschool Zadkine (Watchmaking Vocational School of Zadkine). The hugely successful launch of the manual Heart Beat Manufacture in 2004 was followed by the introduction of the hand-wound Heart Beat Manufacture with Moon Phase and Date in 2005, before the first series of Automatic Heart Beat Manufacture watches was launched in 2006.

Through the cooperative efforts of all concerned, the newest generation of Frédérique Constant watches was turned from an improbable concept into a wonderful reality. Frédérique Constant is proud to cooperate in a spirit of trust and happiness with these outstanding educational institutions.

COOPERATION WITH CHARITABLE ORGANIZATIONS

Because Frédérique Constant has grown in just over two decades to become a leading manufacturer of Swiss watches, we find ourselves in the fortunate position of being able to cooperate with charitable foundations and other worthwhile causes around the world. We have done so with pleasure and pride, and we will continue to cooperate with charities that can help people around the world -- particularly children -- who are in need..

Over the past few years, we have cooperated with luminaries such as Eva Longoria, when Frédérique Constant sponsored a gala hosted by Eva for the Par Cœur Association, which constantly seeks to help children and their families around the globe. We regularly donate to the International Children's Heart Foundation. We also worked in cooperation with the "Paint a Smile" Foundation to transform a ward of the Beijing Children's Hospital, and we have made numerous donations to organizations that work to detect, prevent and cure heart disease.(2)

As a responsible employer, supplier and member of the world community, Frédérique Constant recognizes and welcomes the need for cooperation at every level. We whole-heartedly commit to carrying on in this spirit of cooperation in the fervent hope that through continued growth and success, we can also continue to do good wherever we go.

OLIVIER ZAUGG
Operations Director

OLIVIER ZAUGG ON "COOPERATION"

What led the current Operations Director to Frédérique Constant and Alpina?
OLIVIER: Basically, I was already on my way to Malaysia, where my former employer – a manufacturer of watch and medical mechanical components – had asked me to rejoin the firm. As a supplier for the watch making industry, I naturally knew many of the colleagues in Geneva, and one of them persuaded me not to move to Asia. So I unpacked my bags and began my career in Aletta and Peter Stas' family business.

Did you know what you were getting yourself into?
OLIVIER: Of course. Before accepting their offer, I had carefully researched the special and unique working environment here. Frédérique Constant and Alpina are different from all the other watch making companies. One decides to come and work here! But I must admit it was precisely this different working environment that appealed to me.

In your opinion, what exactly is different here? What makes this company different from others?
OLIVIER: When they're not traveling, the owners spend the whole day here with us. They work hard alongside us and are an active part of the whole. Hierarchies are extremely flat. If you need something from Aletta or Peter Stas, you don't have to make an appointment. All you need is a bit of patience until they can find a suitable moment to talk with you.

What about your colleagues? What impression did they make when you first arrived?
OLIVIER: They impressed me as passionate, dedicated, small businesspeople within the enterprise. They work independently, take plenty of responsibility and do their jobs. This is definitely not a nine-to-five business. We colleagues also talk to one another a lot. We share ideas and mutually inspire one another.

Could you describe the staff at Frédérique Constant and Alpina in a few words?
OLIVIER: Yes! Genuine, enthusiastic about their work, liberal-minded and tolerant.

PRECISION
High technical skills required to assemble by hand

MANUEL GOMES DE ALMEIDA
Uhrmacher
watchmaker

DANIEL AEBERHARD
Uhrmacher
watchmaker

Togetherness in the Frédérique Constant and Alpina extended family even includes the suppliers and independent distributors of both watch brands.
OLIVIER: That's right! We're all in the same boat, so to speak.

How often do you meet with Aletta and Peter Stas to discuss the current situation?
OLIVIER: We discuss the most important items daily. Who's doing what? How do the key performance indicators look? What obstacles are we struggling against or are looming on the horizon? As I see it, the most important thing is work together to constantly keep the extended family's metaphorical boat on course, while sailing close to the wind. If we can do that, then we work very efficiently and enjoy the ensuing success.

Each department head plays an important role here. How often do you see them?
OLIVIER: We meet once a week. Each department head reports on the status of his department and describes the momentary successes or problems. If there are latent problems, we collaboratively search for solutions. Acute problems are handled immediately and without a prior appointment. Our engine must always run smoothly because our goals are flexibility, responsiveness and punctual delivery – earlier than our competitors, whenever possible. If you keep everything onto the right course, if you sail hard against the wind and if you also want to be flexible and responsive, then you've got to stay damned alert and be quite sure that the quality doesn't suffer.

How do you reconcile those opposites?
OLIVIER: Success comes only through optimal planning with all suppliers and our own individual departments. All processes must be in harmony with the high quality that we desire. Ultimately, the quality determines the tempo. Whoever claims otherwise is a fibber or a magician.

Where have hitches been most likely to occur lately?
OLIVIER: You probably won't believe me, but it's the payment behavior of some of our new distributors. All of them ultimately pay, but international cash flow has repeatedly created severe problems. Not every country takes payment morality quite as seriously as we Swiss. But we're an extended family, so the most difficult part can be to treat our debtors with politeness and style. After all, we don't want to affront anyone.

When and how do you relax?
OLIVIER: Theoretically, I'm accessible to the company and active on its behalf 24 hours a day, seven days a week. I enjoy sports in my leisure time. Right now, I'm building a house because I've moved with my family into a home near Geneva. That keeps me pretty busy.

Frédérique Constant's Core Values

PERSONAL DEVELOPMENT

At Frédérique Constant, we recognize that any company's most valuable asset is its workforce. Nowhere is this more true than within our organization, because we actively seek a highly diverse range of talented individuals who must be exceptionally passionate about their craft in order to join our team.

Even though we're all in the same team, all people have different needs, whether socially, domestically or in their educational aspirations. As such, Frédérique Constant lists "Personal Development" as one of its core values. Of course, recognition is one thing, while actions to permit and encourage personal development are a different matter entirely.

FLEXIBILITY TO SUIT DIVERSE NEEDS

Many members of our workforce are eager to continue their personal development through education. Others have family commitments or other circumstances that would make it difficult for them to conduct a "normal" work week. For these reasons, Frédérique Constant has introduced highly flexible work schedules that allow our employees to advance their careers while working either fewer days per week or fewer hours per day.

New parents can take up to a full year's unpaid leave to enjoy those incredible first months of their baby's life in the safe and certain knowledge that should they wish to do so, they may return to their carreer with Frédérique Constant.

We also have a young and dynamic team and we recognize that some people feel they would like an opportunity to travel in order to broaden their minds through diverse and new experiences. While many employers would hesitate to support these desires, Frédérique Constant's employees are allowed to take a sabbatical of up to three month -- not just once, but every seven years they are with the company.

RALPH SIMONS
President
Alpina & Frédérique Constant
USA

NIELS EGGERDING
Adjunct Commercial Director

FELIX DEMARTY
Watchmaker T0

LISE BERROD
Marketing and Communication Executive

JEREMIE HEIDRECHEID
Planning Manager

FROM THE TOP
International Sales Managers
Sebastien Cretegny
Adrienne Ogier-Serat
Williams Besse

PERSONAL DEVELOPMENT THROUGH PARTICIPATION

A dynamic organization will expect and encourage every individual in the company to contribute energy and new ideas in a combined effort to bring even greater corporate achievements. Frédérique Constant places no limits or restrictions on the sources of inspiration which will lead to continuous success. Everyone within the company is welcome to offer suggestions and proposed plans of action that will help make the company bigger and better.

Being an active part of a dynamic and successful company is an excellent avenue toward personal development. By working in an environment that calls on you to use your talents to the fullest of your abilities at all times, you discover a strength within yourself to make a positive contribution, efficiently overcome obstacles, and ultimately contribute to the greater good on both a professional and personal level.

PERSONAL DEVELOPMENT THROUGH TRAVEL

Frédérique Constant watches are sold in over 2,700 outlets in more than 100 countries around the globe. As an international brand, we frequently send members of our sales and management team abroad, where they encounter and interact with widely diverse individuals and cultures.

Through this interaction, our employees gain greater understanding, not just of the complex nature of our business, but of the world in which we live. They also have the opportunity to discover their own strengths and weaknesses, and to develop their personal communication skills while continuing on a successful, professional career path.

PERSONAL DEVELOPMEN THROUGH MEETING THE CHALLENGE EVERY DAY

Anyone looking for a career where everything is rigidly structured and where people just "do their jobs" and nothing more should not consider applying for a position at Frédérique Constant. Our employees are self-starting, self-driven individuals who make success happen rather than waiting for someone else to come up with all the good ideas. This is without doubt a large part of the reason why the company has grown by 25 to 30 percent annually over each of the last ten years.

The motivation of our workforce comes from within; it is not forced upon them. They are passionate about what they do and how they live. And they are the true embodiment of our company's slogan: "Live your passion!"

We ask a lot of our employees, but never more than that of which they are capable. They must be flexible; always ready to adapt and to respond in a positive fashion to the ever-changing nature of the watch industry. They should be curious, always seeking out the next trend, the newest innovation, the "eureka moment" that will drive the company to ever-greater heights.

Our employees need to truly understand all of the individuals who are ultimately a part of the company's success. This includes not only the customers we cherish, but also the company's stakeholders, suppliers and fellow employees.

SYLVIE CRETIN
Receptionist

CHRISTIAN GRANDJEAN
Transport Manager

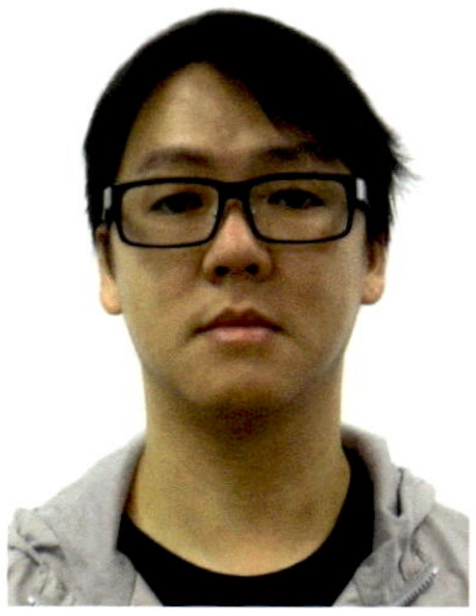

ANDREW WOO
Graphic designer

DORA HO
General Manager Hong Kong

GARY SO
Product Manager Hong Kong

JADWIGA GLASSEY
Accounting Department

PERSONAL DEVELOPMENT THAT LEADS TO SUCCESS

At Frédérique Constant, we believe that by learning and growing together, and applying our skills and knowledge collectively, we will ultimately triumph over our less dynamic rivals in the marketplace. We also believe that we can work together to become the very best in the business, and we resolve never to cease our desire to grow and develop personally and professionally, both as a company and as individuals.

We feel that part of personal development is accepting responsibility and accountability. Every member of the Frédérique Constant team should consider the company's assets as their own, and they are asked to recognize the need to make a contribution to greater corporate efficiency through the improvement of company systems.

In return, we fully commit to showing respect for each and every individual within the company, and we salute the artistry, passion, dedication and loyalty of our fellow employees.

HUGO HERRERO-LEVY
Product Manager

INTERVIEW WITH HUGO HERRERO-LEVY ON "PERSONAL DEVELOPMENT"

Hugo, what structures did you find when you joined the firm in 2008? What prior education had you already had when you joined Frédérique Constant?
HUGO: After graduation from the Art & Métiers engineering school, I worked with some friends in a collaborative consulting firm. But there wasn't much money to be made in it and I yearned for a real career. Joining Frédérique Constant precipitated a long-term change in my life. When I look back, I must admit that my image of the watch industry had been totally different. We'll talk about that later! First, let me return to your question. I was initially hired to head the "T1" atelier. My job was to set up a factory for manufacturing watch movements. The structures I found were simple and not yet sufficiently structured.

How many colleagues were there?
HUGO: Ten. I was 24 years old and everyone else in the group was likewise in their mid twenties. My first task was to persuade people to listen to me. A precise analysis of strengths and weaknesses helped. I also had some doubts about my work at that time because Pim Koeslag had already set up the structures, in accordance with his ideas, for the ateliers where calibers FC 930 and 935 were assembled.

So you couldn't turn everything upside down. Or could you?
HUGO: No, not until we began my own project -- which was to assemble Alpina calibers 650 and 950 -- could I specify the exact structures for the work sequences. Every step from A to Z was planned.

Did you also have to look after collaboration with suppliers?
HUGO: Yes, and that brings me back to what I mentioned at the start. I had imagined that sequences in the watch industry would be somewhat different. Not only the customers, but also the suppliers had their say. That's unconventional. It's found in this form in only a very few industries. Above all, one doesn't learn it in any university. It was essential for me to speedily develop personal relationships with individual suppliers. I had learned that one works in the watch industry with friendly manufacturers and not with suppliers. This branch of the industry is very human; it's ultimately one big family. Nowadays I'm on a first-name basis with most of our suppliers. Frédérique Constant was likewise already a big family and has remained so, although the clan has grown larger.

How does that affect the way you think about and approach your work?
HUGO: After I'd understood how the wind blows, I did everything I could to assure that our people would be eager to

come to work. I gave top priority to creating optimal working conditions and a correspondingly pleasant working atmosphere.

MARIE-PIERRE MORIN
Assistant of production

So it seems that Aletta and Peter Stas gave you good chances to develop yourself and you seized those opportunities. You're production manager now and your responsibilities within the company are growing progressively.

HUGO: That's right! The Frédérique Constant and Alpina brands offer much more than their competitors. Every colleague who wants to move up gets a chance to do so. We never put obstacles in the paths of ambitious employees. There are no waiting lists for jobs and there's no mobbing. We're also not afraid to hire people who haven't already worked in the watchmaking industry. For example, a young mother came to us from the printing industry. She wanted to change her career and to have more flexible working hours. No problem: we made that possible for her. She's presently working as an assistant in logistics and purchasing. As long as each individual is conscious of their duties and does their job conscientiously, we respond in kind by offering flexibility at the workplace. In this young mother's case, it meant that she would have more time for her child. Both sides – the employee and the employer – have remained satisfied and now I can move on to take care of the next case.

Do you also train apprentices?

HUGO: Yes, we are happy to work with trainees and apprentices. Young people are the future of our industry. We must introduce them to the fascination of watchmaking and spark their enthusiasm for what we do. It's wonderful that most trainees who have finished their internships return to work for us. Many of them continue their education here because they want to move up and have a lifetime career with us. Aletta and Peter Stas encourage that as much as they can.

Hugo, later you were responsible for ateliers "T2" and "T3." You were especially able to show what you could do during the recent crisis, when you kept costs low and simultaneously kept productivity and morale high.

HUGO: I learned a lot during those years! The objective was to achieve everything simultaneously. It wasn't easy.

What are you working on right now?

HUGO: Together with Aletta and Peter Stas, I'm presently optimizing the industrialization of our own calibers. That's the focal point of my work and there's still plenty of work left. Our goal is to bring as many steps of the work as possible onto our own premises. If this doesn't make sense or isn't feasible, then we want, at least theoretically, to master each step in the work.

With so much work to do, do you still find time for private life?

HUGO: Private life merges very strongly with the company. We're also reachable outside of working hours. Some things simply cannot wait. When decisions are urgent, we need to make them even if we're on vacation. My wife isn't always thrilled about that. On the other hand, she's happy that we've grown to become a part of this big congenial family.

FREDERIQUE CONSTANT
GENEVE
SWISS MADE

PETER & ALETTA

TALK ABOUT THEIR MANUFACTURE

A watch manufacture in Switzerland may only call itself a manufacture if it has and makes at least one movement of its own. It took you about ten years to evolve Frédérique Constant from an établisseur, i.e. a business that assembles watches from purchased components, into a full-fledged manufacture.

PETER: The manufacture project can be compared to the history of our watches. We talked with scientists and people from the industry, e.g. with a representative of Precitrame, the machine-builder in Tramelan. They've always been very helpful to us. We also spoke with KIF about shock absorbers and the regulating system.

What about the components for the oscillating and escapement system?

PETER: Nivarox-FAR has always been very accommodating with assortiments, i.e. sets consisting of an escape-wheel, pallet, balance, hairspring and accessories.

FIRST MANUFACTURE
Heart Beat Manufacture
FC-910MC4H6 (2004)

ONLY WATCH
Heart Beat Manufacture Moonphase
& Date - Unique Piece
for ONLY WATCH Charity (2008)

ALETTA: We measured the movements and conducted in-depth research. Then we talked with instructors at watchmaking schools to find out if their graduates would be interested in a project of this sort.

So 2004 was decisive?

ALETTA: Oh yes. We were very excited when we presented our own caliber. Hand-wound Heart Beat FC-910-1 was a tremendous leap forward for our Heart Beat philosophy.

PETER: To develop it, we began collaborating with Pim Koeslag, a very gifted graduate of the school of watchmaking in Schoonhoven, Holland. Working closely with him, we inverted a classically constructed movement to reposition the big balance onto the front, where it's optimally visible through an aperture in the dial at "6 o'clock."

This caliber calls to mind a tourbillon movement.

PETER: Exactly. But one needn't necessarily have a tourbillon to see and hear the heartbeat of human culture, i.e. the ticking of a mechanical watch, with one's own eyes and ears. Our Heart Beat manufacture calibers hide none of this fascination.

ALETTA: We also put glass backs on wristwatches that are fitted this caliber.

How long did you work on this movement?

PETER: We began developing the manufacture caliber in 2001. From then on, a crew of eight collaborated with specialists from the school of watchmaking in Geneva, the school of engineering in Geneva and the school of watchmaking in Holland that I mentioned before.

ALETTA: Their job wasn't simple. The objective was to bring the balance and its tiny hairspring as far forward as possible and to position them at a predefined location beneath a circular opening in the dial, where they tirelessly complete 28,800 semi-oscillations per hour.

page 140-142
MANUFACTURE
FC-938MC4H6 (rt)
FC-938CDG4H6 (rb)
Drawing Caliber Bridge (lt),
Assembly of a Heart Beat Manufacture (rt), Atelier T1 (b)

FREDERIQUE CONSTANT GENEVE
SWISS MADE
Frederique Constant - FC935X4H
Heart Beat Manufacture - Limited edition 1884 pieces
FREDERIQUE CONSTANT
ADJUSTED TO HEAT, COLD, ISOCHRONISM
TWENTY-SEVEN (27) JEWELS
FIVE (5) POS
GENEVE

FREDERIQUE CONSTANT
GENEVE
SWISS MADE
FREDERIQUE CONSTANT GENEVE
SWISS MADE

PETER: This demanded a wholly new movement construction from A to Z. Conventional calibers from ETA couldn't be suitably modified.

ALETTA: The team created an exclusive microcosm with a mainspring that stores enough power to keep the balance vibrating for 49 hours, after which time the watch needs manual winding to replenish its reserve of energy.

PETER: Particularly important in this context is that we remained faithful to our motto of accessible luxury and created a genuine manufacture movement that's nonetheless affordably priced.

You didn't stop with only one caliber.
ALETTA: That's right. Caliber FC-915, the moon-phase version from 2005, was followed a year later by our exclusive automatic manufacture movement with alphanumeric designation FC 930.

PETER: We focused on a new and innovative material in 2008. Our self-winding caliber FC-935 Silicium with moon-phase display and hand-type date indicator relies on a specially designed silicium escape-wheel in tandem with a conventional pallet.

What are the advantages of silicium?
PETER: It has extraordinarily smooth surfaces so conventional oil is no longer needed on the escape-wheel's teeth.

Your eagerness to experiment and your courage to take risks didn't stop there. One could say that Frédérique Constant acted in a truly visionary way.
PETER: If you're talking about what we presented under the designation FC-935SZABS4H9 in September 2008, I would agree. It was unique because in addition to its silicium escape-wheel, it also had a totally innovative balance made from a glass-ceramic substance called "Zerodur." I believe this was unprecedented in watchmaking history.

ALETTA: That watch was sold at the "One of a Kind" charity auction in Monte Carlo. *(1)*

Where do you stand today?
PETER: Our manufacture project is progressing step by step. Our vertical range of production currently accounts for about 15% of the value of our own calibers. We annually produce circa 10,000 pieces and we're working to double that quantity.

ALETTA: Brass components in series of 1,000 units or less are made on our own machine in the cellar.

You also offer your own manufacture tourbillons.
PETER: We initially used modified tourbillon calibers from Progress, but that business was sold and renamed STT. As we gained more experience as a manufacture, we became confident that we would also be able to make our own tourbillon.

ALETTA: Our "Heart Beat Tourbillon" debuted in a limited edition of 188 watches in 2008.

PETER: This movement has several special features, e.g. the so-called "Smart Screw" system to perfectly balance the rotating cage. The complex automatic caliber FC-980-1 also has an antimagnetic silicium escape-wheel.

Tell me about the production of this tourbillon. Who supplies you with which components?
ALETTA: We manufacture the basic movements for the tourbillon wristwatch on our own premises. Mimotec delivers the components for the cage and the springs. Sigatec supplies the pallet and the escape-wheel. Between 2001 and 2011, we systematically acquired experience in manufacture work, i.e. the fabrication of components and the assembly of complete movements.

PETER: We're also progressing towards full autonomy in pallets, escape-wheels and balances. We're still in the R&D phase, but we've already come a long way. Together with specialists, we can now fabricate staffs for the pallet, escape-wheel and balance, as well as the entire balance.

ALETTA: But we don't yet do that in-house.

PETER: As far as the details go, we rely as much as possible on standard assortment 20.3. There are basically three options for the pallet. First, a classical pallet using pallet-stones affixed with lacquer, but that's extremely difficult for us, so it's only a last resort. Second, a silicium pallet based on a tried-and-tested technology.

ALETTA: We've used a silicium pallet in our tourbillon since 2009 and it functions very well. It's somewhat costly, but not astronomically expensive.

PETER: Third and finally, we had the idea of manufacturing the pallet, including its fork and staff, from a single solid block of ceramic. We registered this concept for patent protection in Switzerland in 2011. The first prototypes are now available. If this system functions as well as we hope, we'll be able to use it instead of the corresponding components in assortment 20.3. *(1)*

What about the prices?
PETER: We expect the price of the new serially produced ceramic pallet to be lower than the cost of a classical Nivarox pallet. We're truly competitive there.

ALETTA: We could post-process with diamond abrasives, but we hope that won't be necessary. The manufacturing tolerance is one 1,000th of a millimeter and the edges are extremely sharp.

(1)

Frédérique Constant supported the ONLY WATCH charitable auction in Monaco

(1)

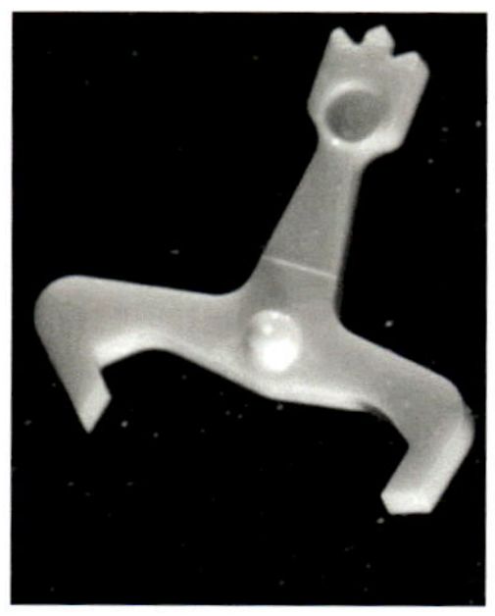

CERAMIC ANCRE

TOURBILLON
Heart Beat Manufacture Tourbillon with Silicium escapement wheel FC-980MC4H9 (2008)
Limited edition of 188 pieces

FREDERIQUE CONSTANT
GENEVE
SWISS MADE

FREDERIQUE CONSTANT
GENEVE
SWISS
MADE

PETER: For the depth of meshing, we can refer to silicium technology, where 80% of the pallets on each wafer have classical depth of meshing. The remaining 20% are distributed among four different types of pallets, each with its own depth of meshing. If we encounter any problems, we could simply use a different pallet.

What advantages does this technology offer for you and your customers?

PETER: We expect that the ceramic escapement, like its silicium counterpart, will enable us to eliminate the need for lubrication. We've found that the amplitude of the balance's oscillations is much higher and more stable when we use silicium rather than the conventional assortiment.

What about the balances?

ALETTA: Peter and I attended the suppliers' fair in Moutier, where more than 450 exhibitors were represented, e.g. tool- and machine-makers, as well as suppliers of cleansing and polishing agents. Toward the end of the fair, we found a company that also mills balance-staffs. We introduced ourselves and explained that we were looking for a manufacturer of balances.

PETER: After we had discussed the number of units and the purpose which we had in mind for our caliber, they asked us for a bit of time to learn more about our company and to consider our proposition. In the meantime, we've progressed far enough to send our balance's design to them so they can respond with a cost estimate.

ALETTA: One of our strengths is that Peter and I are often personally present. Our ambition is to continue learning and developing ourselves, which ultimately benefits Frédérique Constant.

CLASSICS MANUFACTURE
FC-710MC4H4 (2012)

CLASSICS MANUFACTURE
FC-710MC4H4 (2012)

FREDERIQUE CONSTANT GENEVE
SWISS
MADE

We haven't yet spoken about the controversial theme of hairsprings. No mechanical movement can run without one. I suppose Frédérique Constant too is still strongly dependent on Nivarox-FAR?

PETER: Unfortunately, yes, but a good alternative is gradually growing with Haas in Germany. We're currently negotiating our first order with Haas and then we'll see how good the quality is, but I'm quite confident because both Haas and Nivarox use the same material.

ALETTA: Haas supplies all types of hairsprings, either with the blanks alone or with hairsprings that have collets, have been counted and have received their terminal curvature – whatever we need and would like.

But each balance must always perfectly match its hairspring and vice versa.

PETER: Yes. We match hairsprings and balances on our premises. We were very fortunate to find an Omega-Metric. The machine is no longer made, but it's indispensable for matching balances and their hairsprings. Almost every Omega-Metric disappeared from the market in the wake of the Quartz Crisis.

ALETTA: We were quite lucky to contact an older gentleman who had discovered five used Omega-Metric machines in the Baltic countries. The old electronics were removed and new computer-controlled electronics were installed in Switzerland, but the basic hardware remained unchanged.

PETER: We had tremendously good fortune there because four of those five machines had already been sold to Renaud & Papi, Rolex and Atokalpa. Only one – and definitely the last one – was sold to us.

Who at Frédérique Constant knows how to use that machine?

ALETTA: The seller will now teach us how to use the machine properly and efficiently. Theoretically, we should be able to process 5,000 hairsprings each week, which would add up to about 240,000 balance-springs per year.

TOURBILLON MOONPHASE
Heart Beat Manufacture
Tourbillon Moonphase and Date
FC-985MC4H9 (2010)
Limited edition of 188 pieces

Balances are like automobile tires. They must be perfectly balanced.

PETER: We acquired a Greiner Balance-O-Matic for precisely that purpose. We had to invest a rather large sum of money for both machines, but our investment will make us more independent.

ALETTA: Together, the two machines are like an insurance policy. If something really severe should happen to deliveries of balances and hairsprings, we would be able to continue producing our manufacture calibers without outside assistance.

PETER: If everything goes as planned – and that, of course, is our goal – then Frédérique Constant will sooner or later achieve genuine autonomy in the assortiments sector.

Are you planning to launch other complications based on the Heart Beat caliber?

PETER: We ceased collaboration with Dubois-Dépraz and other extramural developers eight years ago, and we've worked entirely in-house ever since.

ALETTA: It soon became clear that the in-house alternative is much less costly.

PETER: We've presented at least one new movement each year since 2004. The Heart Beat caliber came first and then we launched our "Maxime" as a basic movement. Both were soon augmented with additional functions. The Maxime was given a centrally axial seconds-hand in 2012. Of course, that wasn't as spectacular as our "Worldtimer" with its readily settable and easily adjustable 24-hour ring and city ring.

If you'll pardon my pun, what do you have up your sleeve for the coming years?

PETER: A chronograph, for example. It may be difficult to believe, but a caliber with a stopwatch function is a very complex affair. We had initially planned to present it in 2013, but we've now realized that it will take more time, so we revised the scheduling and are now aiming for a market launch in 2014. We're also working on a movement with a moon-phase display in a very classical style.

WORLDTIMER
Classics Manufacture Worldtimer
FC-718WM4H6 (2012)
Limited edition of 1888 pieces

FREDERIQUE CONSTANT
GENEVE
NEW YORK
RIO DE JANEIRO
DENVER
CHICAGO
PUERTO RICO
S.GEORGIA
AZORES

FREDERIQUE CONSTANT ★ MANUFACTURE
SWISS MADE
26 JEWELS
FC 700 / 720
FC718X4H6
Frederique Constant
Swiss Made
Sapphire - Ltd ed 1888 pieces

FREDERIQUE CONSTANT
GENEVE
KARACHI
BANGKOK
DHAKA
MOSCOW
DUBAI
HONGKONG
CAIRO
SYDNEY
GENEVA +1
GMT
LONDON
-1 AZORES
S.GEORGIA
RIO DE JANEIRO
PUERTO RICO
NEW YORK
CHICAGO
DENVER
LOS ANGELES
ANCHORAGE

ALETTA: We've got quite a few exciting new items in our the pipeline for 2014. Among other newcomers, we're planning to launch a special GMT function.

How much time do you expect the developmental process to take?

PETER: It varies depending on the specific project. If we're talking only about a module, then we anticipate about a year or 18 months at most. We've developed our own processes. Fundamental development typically takes about three months. Then we build functional prototypes as speedily as possible with the aid of our own machines.

ALETTA: Only with such prototypes can we move ahead. If something doesn't work properly, we optimize it on the computer so we can immediately produce the modified components here on our premises. This learning curve enables us to move very rapidly through the three phases of the prototype stage.

PETER: We're not talking about a basic caliber here. That usually consumes about three years.

Does Frédérique Constant still work with Chronofiable?

PETER: Not in every instance, because that adds another six months to the equation, but we do send our basic calibers to undergo the evaluation process so we can avoid encountering difficulties later on. We definitely owe that to our clientele.

ALETTA: The movements that we make in our own manufacture mustn't be one iota less excellent than the ones we purchase from our extramural suppliers.

PETER: Quite the opposite. With our own calibers, we offer exclusive, highly reliable and precise mechanisms to watch aficionados who are demanding and discriminating, but simultaneously also price conscious.

WORLDTIMER
Classics Manufacture Worldtimer
FC-718WM4H6 (2012)
Limited edition of 1888 pieces

KARACHI
DUBAI
DHAKA
BANGKOK
MOSCOW
CAIRO
HONGKONG
TOKYO
GENEVA +1
GMT
LONDON
SYDNEY
-1 AZORES
S.GEORGIA
NOUMEA
RIO DE JANEIRO
PUERTO RICO
NEW YORK
CHICAGO
DENVER
LOS ANGELES
ANCHORAGE
GENEVE
MADE
SWISS

THE GENESIS OF A WRISTWATCH:

(1)

Measuring Balance Wheels under the microscope

(2)

Quality Check of Escapement Wheels

DEVELOPMENT
Silicium Escapement Wheel

FROM MANUFACTURE CALIBER TO COMPLETE TIMEPIECE

Everything turns, everything moves. The crowning creation for any watch manufacturer is to built his own movement, freely and in accordance with the motto: "Live your passion!"

Have you ever asked yourself what's really so exciting about a mechanical watch? In our opinion, it's the movement! We reached this conclusion not because we choose not to peer beyond the nearby horizon circumscribed by the case's rim, but because our passion isn't only to sell wristwatches, but also to develop movements of our own. The mechanical microcosm under the dial, to which we would like to devote some thought today, performs incredible feats. At the very least, this little machine – a Lilliputian confection of bridges, springs, levers, cocks, screws, pinions and gears – tirelessly keeps us informed about the time of day or night. Depending on the version, it may also keep us posted on the date, day of the week, lunar phase and many other time-related facts.

The heart of a mechanical watch is its escapement, which subdivides the linear flow of energy that it receives from the barrel and thus makes time display possible. The hairspring and balance-wheel together comprise the balance of the escapement. Frédérique Constant usually relies on balances that complete 28,800 semi-oscillations per hour and are accordingly paced at four hertz. This results in 691,200 semi-oscillations per day and 232,243,200 semi-oscillations per year. Here's that number written in words rather than digits: two-hundred-million two-hundred-forty-three-(thousand) two-hundred. That's quite impressive! Let's cite another superlative. One day consists of 86,400 seconds. If a wristwatch gains or loses four seconds per day (although such a large inaccuracy is unusual nowadays, then its daily deviations typically remain within even narrower limits and the watch keeps time with a precision of 99.995%, i.e. very nearly 100%. That's impressive: it sparks our enthusiasm, and it fuels our passion! (1)

But the balance is only one of many components in a movement. Gears or pinions likewise rotate many times each day and yearly traverse – relative to their small size – an incredibly long distance of several kilometers. But this microcosm isn't solely inhabited by marathon runners, because other gears or pinions are comparatively lethargic: they complete a full rotation only once a month, once a year, or – in the extreme case – once every four years. But when their big moment finally arrives, they function properly and perform the actions for which they are intended. Levers and springs act similarly: they reliably dispatch their predefined tasks. To cut a long story short, everything moves, everything turns, and everything does so with incomparable precision. (2)

Peter neatly summarizes all this when he says, "Building our own movements was neither obligatory nor optional for us, but the fulfillment of a long-cherished dream. It simply means that we live our motto: "Live your passion!" An earlier expression of it was the opening that we cut in the dial so the indefatigably swinging balance would have a worthy stage. An aperture through which one could admire the balance's choreography was an early expression of our dedication to the fascinating world of watchmaking.

Some people who wear mechanical wristwatches unfortunately realize much too rarely that their wrists are theaters where miniscule components dance a perfectly choreographed ballet. Watch owners are all too likely to treat this microcosm without due consideration or proper respect. Shocks, blows and exposure to heat and cold are a few of the thousand natural shocks to which a wristwatch is daily subjected. With this in mind, we really ought to treat our faithful ticking companions with more respect. We would also be wise to devote a few of each day's 86,400 seconds to the conscious contemplation of our wrist-borne pals. Devote? Yes, indeed! Let's simply take a moment – or more! – to enjoy our wristwatches and consciously realize what's actually happening under their dials. Freely and in accordance with the motto: "Enjoy our passion!"

To design and manufacture a movement and to produce a completed wristwatch really isn't witchcraft. You simply have to do it. On the following pages, we invite you to accompany us along the path which leads from the initial idea and the first steps, through preassembly, final assembly and insertion of the movement into the case, and ultimately to the final quality control.

The first impulses usually come from the design department, where the idea for a product occupies the focal point. A new movement with a mechanism for a time-related parameter will need to be built so that a watch with a corresponding display can be created. The project director and design engineer necessarily participate in the discussions when the time comes to begin talking about the need to make a new movement. Their analysis determines whether the movement will have to be constructed entirely anew or if an existing time-honored caliber can be modified. If necessary, the technicians can also check the overall feasibility of the project. (1)

Money also plays an important role. Aletta, this probably doesn't have anything to do with miserliness, or does it? "We at Frédérique Constant and Alpina want to democratize luxury and make it accessible to more people. Careful calculations invariably precede the decision to bring a new line of watches or a new model to the market." The expenses must remain within the limits specified for the project. Practically anyone can produce a fine wristwatch by spending a king's ransom, but it's far more difficult to find solutions that neither cut corners on refinement nor cost astronomical sums. The designers, engineers and watchmakers accordingly always strive to find the best mix of quality, costs and the desired scheduling for completion and delivery. Peter, cost-optimized working, followed by punctual delivery, that's your specialty, isn't it? "Of course, it pleases Aletta and me, as well as all the others who are involved in the genesis, and it ultimately benefits our dealers and customers."

But we're getting ahead of our story. Before we continue, we need step back and explain some basic concepts. The watch industry divides the process leading toward a completed watch into four phases from "T0" to "T3." Drawing the initial constructive sketches and evaluating the feasibility occurs during "T0," which also includes the making of prototypes, the manufacturing of the zero series and the industrialization. This is followed by "T1," when individual components needed for the assembly are made and prepared; the movement is also assembled and finely adjusted in this stage. "T2" consists of installing the dial and the hands, as well as encasing the movement. The wristband is attached during "T3" and engravings (if any) are added to the case: certificates (if any) are issued, and the completed watches are packed for shipment.

The genesis of a watch movement needs more than ideas and passion. Foresight and the courage to invest large sums of money also play an essential role. The integration of phases "T0" through "T3" under one roof at Frédérique Constant and Alpina was only possible after our relocation to a spacious industrial building in Geneva's Plan-les-Ouates. Subsequent acquisition of computer-guided multi-axis processing centers (CNC) and an appropriate machine park further enabled Frédérique Constant and Alpina to design a watch movement and monitor its genesis from the earliest conception to its final completion.

Now let's return to the idea and the planning. After the 3D plans with all necessary details have been drafted in the construction department, the work leading to fabrication of the first prototypes begins in the light-flooded cellar ateliers in Plan-les-Ouates.

(1)

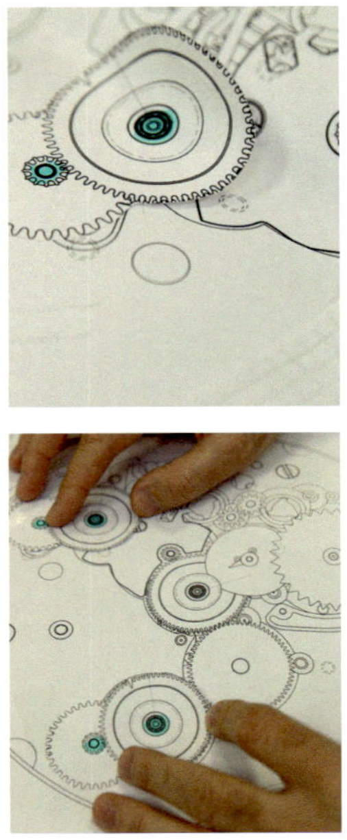

Developing a new caliber, where to place wheels

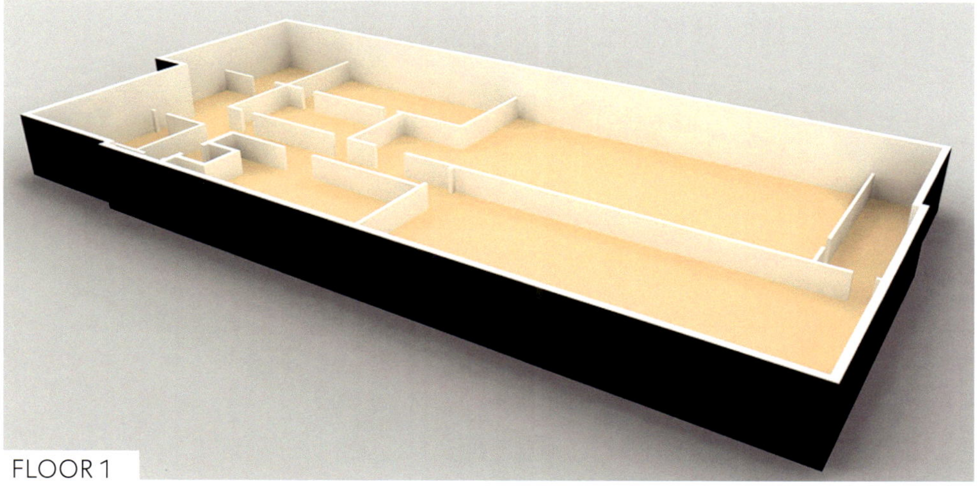

FLOOR 1

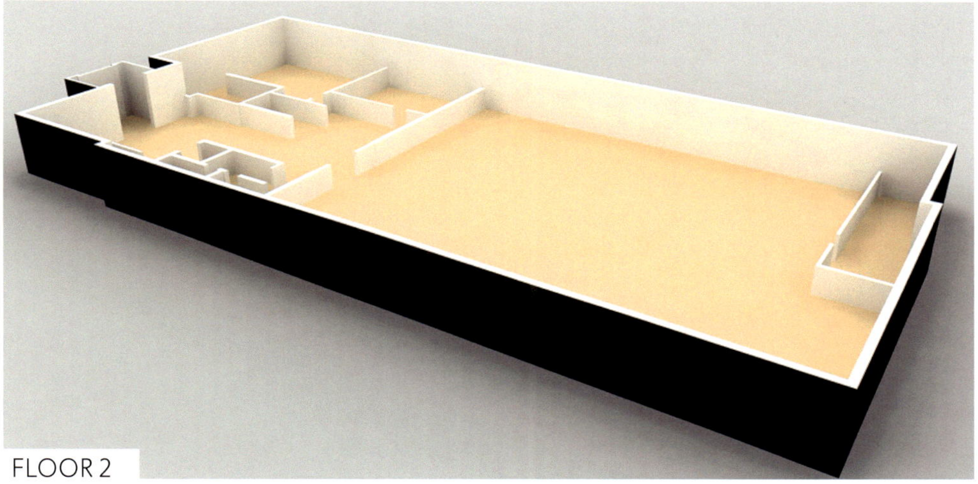

FLOOR 2

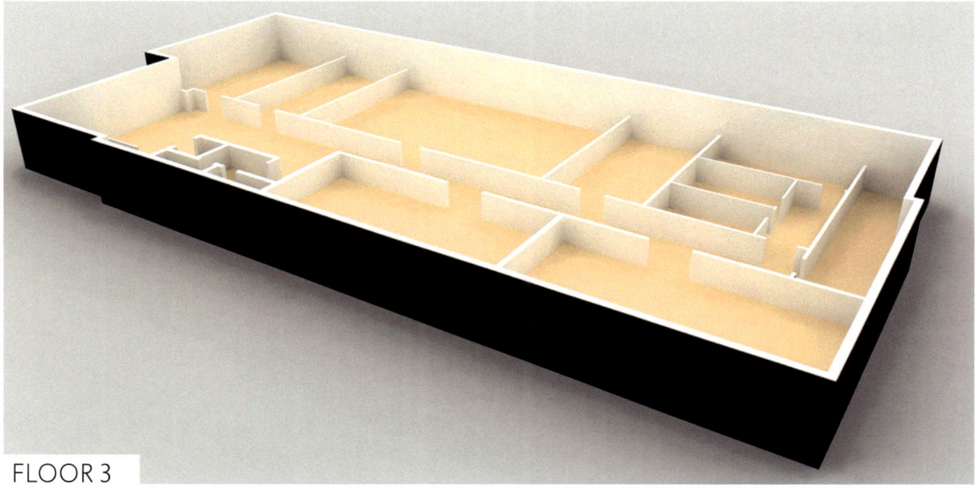

FLOOR 3

MANUFACTURE
Manufacture in Plan-les-Ouates
3D view of the different floors

TO ATELIER
Production of the manufacture caliber's components on CNC Machine

LIGHT-FLOODED CELLAR ATELIERS

Plates, bridges, springs, levers, cocks and other small parts for the movement are manufactured piece by piece and with the aid of CNC machines. After a CNC machine has been programmed, it can autonomously mill, drill and cut. It also automatically changes tools for the various tasks and monitors the amount of wear suffered by these tools. Whenever necessary, the machine pauses, extracts the next drill bit or milling tool from its toolkit and then resumes its robotic labor. Meanwhile, external partners and suppliers are commissioned to fabricate screws, wheels and pinions for the soon-to-be-built prototypes. This is also the phase in which the design engineer and project director communicate with their in-house colleagues in the prototype department and with external suppliers. Even the tiniest details are relevant, and some of them require post-processing to tolerances in the hundredths- or even thousandths-of-a-millimeter range. A reconstruction or redesign is seldom perfect right from the start, so all participants must work to the best of their abilities during these crucial days, weeks and months. A golden rule of construction states that one should intensively use the conceptual phase, when iterations can be realized much faster and at far lower expense. Plans can be modified and engineers can make the required modifications on the basis of these findings and other quality analyses. Additional prototypes are made and analyzed. Meanwhile, the project director keeps a sharp eye on the schedule and hosts visits from suppliers; the design engineer checks the final details in the sketches and rectifies any remaining tolerance errors, and the designer successively provides the releases. This is followed by plans for the manufacturing and checking. Last but not least comes the homologation. Empirical findings from prototype construction are integrated into the production of assembly tools, after which the manufacturing of the zero series can begin. This zero series corresponds to the industrialization of the movement, a process that's often underestimated by people who haven't fully envisioned its complexity. After having reached this stage, one prays that no further improvements will be needed because every change made during this later phase costs more time and money than it would have during the prototype and zero-series stage.

After some pre-run time has elapsed, and after the movement's components have been made and/or delivered in sufficient quantities, phase "T1" can begin. This involves the manufacture and preparation of individual components needed for assembly, as well as the assembly and fine adjustment of the movement. Before this can begin, components that were made on the premises and other parts that have arrived from third party suppliers must pass quality-control tests. Based on prior experience, these tests involve either spot checks or scrutiny of every part.

ATELIER TO
Tools for CNC (l), Programming the CNC (m), CNC in operation (l)

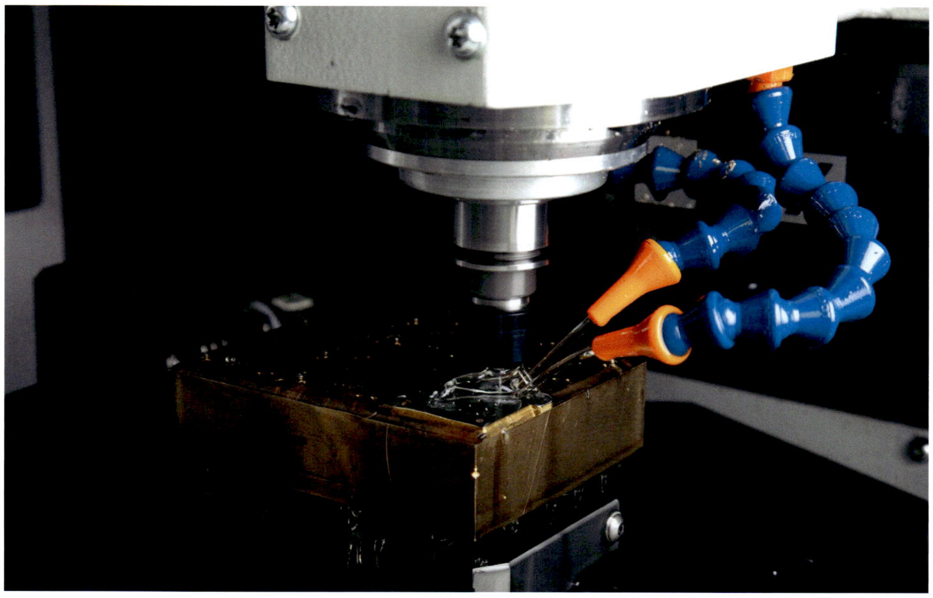

ATELIER T1
Main view (t), Assembly
Tourbillon Caliber (lb and rb)

ATELIER T1

The head of atelier "T1" assures that the tasks are efficiently delegated. Some employees preassemble individual components. For example, they may install the self-winding system, the hand-setting and winding mechanism or an additional function such as a moon-phase display. Others assemble the complete movement. The training given to workers in atelier "T1" varies depending on the degree of difficulty of the tasks entrusted to them. The spectrum ranges from watchmaker's assistants to fully trained master watchmakers. The atelier's manager makes sure that colleagues who have received equivalent training regularly switch from one task to another at various workplaces. This creates greater autonomy and improves morale. Frédérique Constant and Alpina are currently approaching a critical dimension which will justify semi-automated assembly lines for movement assembly. Rather than distributing the individual tasks among various workplaces and then passing the semi-finished pieces onward to the next colleague who'll perform the subsequent manipulations, a more highly automated facility would circulate the movements on a conveyor belt, along which the various steps in the work could be sequentially performed. How does this work in actual practice? At the beginning, the movement plates are affixed in movement-holder rings and placed inside a container. Tasks are sequentially performed at the various workplaces, which are situated one after the other along the line. Each colleague has the necessary components for the movement within easy reach and repetitively performs the same series of tasks. With assistance from the manual assembly apparatus, plates are individually lifted from the worktable so they can be processed. After this work is done, the plate is lowered onto the belt, which conveys it to the next workplace. How can one imagine this in concrete terms? Let's consider, for example, the process of assembling a barrel, which is first pre-installed on its plate and then sent to an adjacent colleague who installs the gear-train. The next step is to monitor the amount of play in the wheels. The gear-train bridge is then affixed, the plate is sent onward and the barrel is installed. Meanwhile, the barrel-bridge is pre-installed by a colleague at an external workplace, i.e. at a workbench that's not part of the assembly line. This is the next part to be installed along the assembly line. Some parts of the barrel-bridge must be lubricated, so this too is accomplished now. The plate is then conveyed to the next colleague, who inserts the pawl and the mainspring. The time required at each workplace depends on the degree of difficulty of the operations performed and on the experience of the colleague. The advantages of this mode of working are obvious. Each worker always performs individual operations under identical conditions. The repetition of identical tasks increases the quality of the work and simultaneously reduces the time required. Time consuming positioning of the plate so it can be optimally processed is unnecessary because the conveyor belt always presents the plate in the optimal form and orientation. Manufacturing can proceed with optimal efficiency after the team on the assembly line has acquired a certain amount of experience and routine. The manufacturer of the conveyor system summarizes its benefits on his website: "The manual assembly apparatus increases the efficiency of the individual steps in the work because it minimizes repeated movements. The user has both hands free to perform assembly tasks per se, while the apparatus automatically conveys individual components to the next colleague." (1)

Movements that have been completely assembled are also finely adjusted in department "T1." The supplier has already coarsely pre-adjusted the escapement, but this must now be improved upon in each of six predetermined positions: dial up, dial down, and crown at "9," "6," "3" and "12 o'clock." It takes a skilled and experienced artisan to finely adjust a watch movement. The art is to find the best compromise between optimal amplitude and an average rate result in the "plus" (i.e. gain) values. Various paths lead to success.

(1)

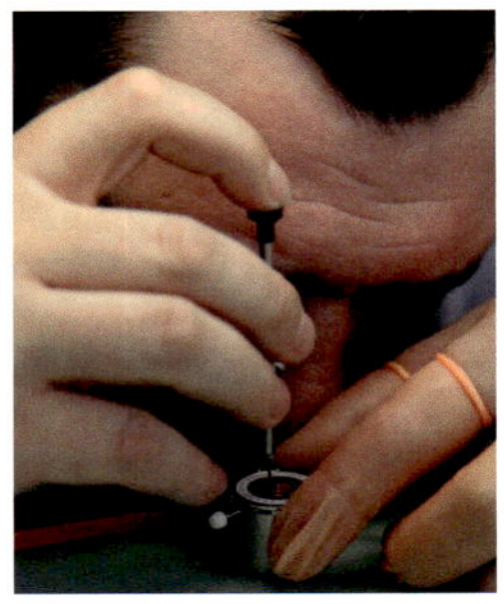

Handmontage eines Kalibers
Assembly by hand of a caliber

Specific standardized interventions in the Swiss lever escapement usually suffice, but sometimes more drastic operations may be needed, e.g. repositioning the pallet-stones. The team in department "T1" also performs another important task, because this is where any errors in construction and planning first become apparent. The atelier's head must be able to analyze any problems quickly and accurately so he can then explain the issue to the project director in the movement manufacturing department. The sooner an error is discovered and rectified, the better. As we explained above, each unavoidable intervention in the ongoing manufacturing process costs an immense amount of money and can indefinitely delay completion of the products.

TOURBILLON
Reglage of the Cage

ATELIER T2

Now let's visit atelier "T2," where dials and hands are installed and movements are inserted into cases.

At Frédérique Constant and Alpina, sets of ten fully assembled and finely adjusted movements are placed in plastic boxes for safe storage and convenient transport from one in-house workplace to another. Atelier "T2" receives these containers and begins to work systematically with each box's tenfold content. Dials and hands are installed first. This may sound like a simple task, but it's easier said than done. The artisan who performs this exacting work must assure that all parts have sufficient vertical play so that none of the watch's hands, in any position, ever touches another hand and/or the dial. Along with enormous manual dexterity, this job also demands a high degree of responsibility. After the hands have been installed, the crown is manually turned to propel each hand through 24 hours. Two orbits around the dial are necessary in order to verify the correct advance of the date display, which should switch to show the next day's date a few minutes before or after midnight. The next step is to insert the movement, along with the dial and hands, into the empty case, which was assembled at a different workplace. The usually three part cases each consist of a middle piece, a bezel and a back. The crystal is now installed either by gluing or pressing it into the bezel. The bezel is pressed into the middle piece of the case. A close-fitting insulator is simultaneously inserted to assure the desired water resistance. To unite the case with the movement, the latter is affixed inside the former with the aid of the movement holder ring. The crown and its stem are installed, thus forming the future connection between the movement and the outer world. If the watch is a chronograph, the start/stop and zero-return buttons have already been installed in the case. A physical connection between the button and the movement mechanically transfers the button's motion to the movement and thus triggers the desired functions, i.e. stop, start and return to zero. The desired degree of water resistance is achieved by inserting O-shaped washers to insulate the tubes for the crown and the chronograph's buttons. Finally, the back is screwed to the case. Depending on the version, this is accomplished either by turning the entire back, which has threads around its periphery, into the correspondingly threaded opening in the rear of the case, or by inserting six small screws through the back and into threaded holes which have been pre-drilled in the case. A closely fitting case-back insulator assures the necessary watertightness. Voilà: a watch is born!

But the timepiece still lacks its wristband and its engravings, if it hasn't already received these engravings in previous stages. Engravings aren't solely embellishments, but also include serial numbers, reference numbers and a specification of the limitation, if the timepiece belongs to a limited series. Attentive readers will no doubt have realized that we've now entered phase "T3," which includes attaching the wristband and engraving the case, issuing any necessary certificates and packaging the completed watches. Nothing but a transparent glass wall separates ateliers "T2" and "T3"at Frédérique Constant and Alpina.

ATELIER T2
View from top and side

ATELIER T2
Assembly of the FC-365B4C6 (2010)

ATELIER T3

Industrious activity invariably prevails in "T3," where all completed watches are thoroughly tested. The movements and cases, to which the wristbands have not yet been attached, are inserted into a Cyclotest apparatus, where they undergo several hours of testing. Each timepiece is rotated around its own axis, while the box is simultaneously carried along a circular orbit like a carousel. These motions simulate the varying conditions and changing orientations which the watches will later undergo on their wearers' wrists and also assure that automatic calibers have ample opportunity to fully wind their mainsprings. The Cyclotest also ferrets out any loose parts. This is essential because on rare occasions, a screw may not have been fully tightened or a component may have been incorrectly affixed during the assembly process. Even if all previous monitoring hadn't discovered a shortcoming, the Cyclotest almost always reveals it. The machines tireless rotary motions loosen any poorly affixed parts, which then tumble noisily inside the case like the proverbial "loose screw." Now is the proper moment for another thorough inspection of the assembled case and its movement. After the timepieces have undergone this simulation of real-world conditions, they're placed in an electronic timing machine which can simultaneously analyze up to ten watches.

The machine automatically turns the cases and movements into the desired positions, i.e. dial up, dial down, and crown at "9," "6," "3" and "12 o'clock." A microphone at each measuring location conveys the ticking of the watch to the computer. The analytical and measuring program draws conclusions, precisely calculates deviations from perfect timekeeping and displays the entire scene on a monitor. Colleagues in atelier "T3" can see instantly whether a tested watch is performing within a specified range. Green numbers confirm good rate values, but a red background or red numbers show that post-processing is needed. The more closely each link in the quality-control chain follows its predecessor, the better the ultimate quality.

The next job is to test each watch's water resistance. Depending on the type of timepiece, this water resistance test may be performed either with a low-pressure apparatus, in which the watch's case doesn't come into contact with water, or via a pressure test conducted underwater. The first-mentioned test is usually sufficient; the second variety of test is primarily reserved for diver's watches. The low-pressure apparatus creates a vacuum under a bell, inside of which the tested watch has been placed. If air escapes from the case, the examiner knows that it is not hermetically sealed. The direction in which the air escapes makes little difference. If the low-pressure apparatus has discovered that a case is not hermetically sealed, water will eventually find its way into the case, where it will cause serious damage. Diver's watches are first tested with the low-pressure apparatus and afterwards tested underwater. For this ordeal, the cases are immersed in a water-filled basin inside the testing device, which is then sealed so that the desired pressure can be exerted inside the apparatus. This simulates the amount of pressure exerted by water at various depths. If the insulating washers fail completely, the results are instantly evident: the case is full of water. Slight penetration by water can be discovered by warming the case and then allowing droplets of cold water to fall onto the crystal: if any moisture has penetrated, the underside of the crystal will fog up, thus revealing that this watch has problems with its water resistance. (1)

The next step is to attach the wristband. This phase truly separates the wheat from the chaff. Merciless visual control under a magnifying glass discovers every blemish, no matter how small it may be. The examiner also determines if any dust has penetrated between the crystal and the dial and/or if the hands or dial are damaged. Regardless of whether it's made of steel, a blend of steel-gold or just gold, each case is meticulously examined for imperfections. The scrutiny also includes the leather wristband or the metal bracelet and its accompanying clasp. If the examiner gives the green

ATELIER T3
Overview (t), QC at work (lb), Rotation Check (rb)

(1)

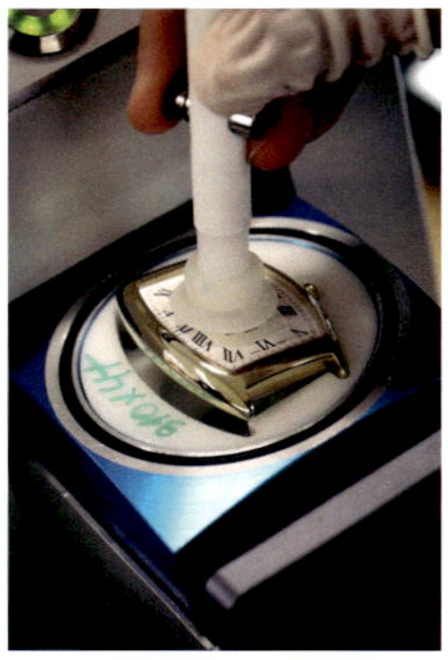

Water Resistance check in the Deltarox tester

(2)

Straps before assembly

light, the finished Frédérique Constant or Alpina wristwatch is placed inside its package. Depending on the order, it will then be sent either directly to the customer or to the merchandise inventory in Plan-les-Ouates. (2)

A noteworthy feature in the genesis of a wristwatch is the large percentage of manual labor involved. Even in today's high-tech era, the majority of steps in the work process cannot be performed entirely by machines, no matter how costly or complex they may be. Nothing can replace human beings, their experience and judgment, their well-trained eyes and unique skills. A team of eighty people work in the developmental and manufacturing departments at Frédérique Constant and Alpina, where they collaborate to create timepieces which will serve as our reliable companions in daily life. Year after year, a wristwatch not only tells us the time, but also gives us so much pleasure because it's a part of us and it faithfully accompanies us through thick and thin. A good accessible mechanical Swiss watch – exactly as we imagine it – fully reflects our motto: "Live your passion!"

FREDERIQUE CONSTANT
GENEVE

ALETTA AND PETER TALK ABOUT

SPONSORING AND CHARITY

Sponsoring and charity enjoy high priority at Frédérique Constant. You started with sponsoring, so let's begin by talking about that.

PETER: In 2004, when our business was still located in Chêne-Bourg, we held a meeting with our advisory board.

Who were the board's members?

ALETTA: Peter, Hans van Willige, a friend from McKinsey, Wim van Heijst, a retired businessman and myself. It was a small group and, incidentally, all Dutch.

PETER: We were working at that time to define the values of our watch brand more clearly. What does Frédérique Constant stand for? Where do we want to go? What are our goals? And what makes Frédérique Constant different from other brands? We realized right from the start that market-sensitive prices rank amongst our most important competitive advantages..

ALETTA: Attractive prices were decisive as a means of reaching younger target groups composed of 30 to 45-year-olds with smaller budgets. Most of these potential customers are business people at the beginning of their careers. Many haven't yet moved up the ladder and are still working in lower-tier executive positions. Their incomes are limited and they may also be starting families, which likewise requires funds.

LADY LOVE HEART BEAT
FC-310LHBPD4

"Let more people enjoy luxury"

PETER: But these young people are nonetheless enthusiastic about Swiss mechanical watches. They're precisely the individuals for whom Frédérique Constant offers attractive timepieces. So we decided that our company should produce accessible luxury in the form of beautiful, affordably priced Swiss watches from Geneva. Our purpose is to enable more people to enjoy luxury.

ALETTA: But beautiful watches at accessible prices weren't enough. We also needed to emphasize the luxury aspect more strongly. So our discussion turned to what we could do to give Frédérique Constant the desired image. That's how we arrived at the idea of sponsoring.

How did you come up with the idea of classic car rallies?
ALETTA: A young Dutchman was working in our communications department at the time and he contributed two good ideas.

PETER: One idea involved balloon flights with a big, dark green balloon prominently labeled "Frédérique Constant." His other idea had to do with rallies for classic cars.

ALETTA: We discussed the suggestions and concluded that car rallies would make sense for us.

Why is that?
PETER: Well, cars and watches simply belong together a priori. Their mechanisms share certain interesting common denominators. That's particularly true for classic automobiles, which needn't necessarily be old timers. They clearly reflect our brand's values.

What did the advisory board decide?
ALETTA: We decided to move ahead with the idea. We delegated the tasks and began to explore various options.

PETER: It was in this context that we found Healey.

ALETTA: A series of telephone calls led us to a Healey club that organizes a meeting in St. Moritz every two years. But we were disappointed to learn that they already had a sponsor in the watch industry.

PETER: They were nonetheless willing to talk with us because they weren't entirely satisfied with their existing partnership. That's how we were able to launch a limited edition of wristwatches for the upcoming event.

ALETTA: We were also very pleased that we didn't need to spend a huge sum of money. We very quickly manufactured various watches as prizes.

How many did you make?
ALETTA: Not so many. If I remember correctly, we made five timepieces. We also delivered polo shirts and caps. All in all, we supplied articles worth about 50,000 Swiss francs.

So your sponsorship wasn't purely altruistic?
PETER: Naturally, we talked with the organizers about a license for a series of watches bearing the Healey logo, but not the Swiss club's logo because, as I said, they already had a sponsor. In that way, we came into personal contact with the Healey Family, which was much more valuable for us. We and the Healeys arrived at an initial agreement on a five-year license.

ALETTA: That agreement enabled us to present our brand at various events and to show our watches in the hotels.

Do you own a Healey?
PETER: No, but Aletta and I took part in a rented car. Our participation was very successful, so we asked ourselves, "What else can we do?"

ALETTA: We joined in another Healey event the following year, which coincided with the 25th anniversary of the Dutch Healey club. The festivities took place in a palace. And we were again present there with our watches.

PETER: That's also true for the planned Healey Museum.

Another success?
PETER: Yes, absolutely. The next year brought the Healey Challenge in Le Mans. Seventy Healeys rolled up to the starting line in the context of the Le Mans Classic. It was the largest fleet of such automobiles ever assembled.

HEALEY
Limited Edition at the Healey event in USA (2011)

HEALEY
Brand partnership with Healey since 2004

FALCON STADIUM
HEALEY
A949P

VINTAGE RALLY
Official Timekeeper of La Carrera Panamericana 2011

(1)

Peking to Paris logo and rotor for the limited edition Chronograph

(2)

VINTAGE RALLY
Limited Edition
La Carrera Panamericana 2011

HEALEY
Concours d'Elegance at Healey USA event (2011) (t), Austin Healey 3000 MKII (b)

But everyone knows that other watch brands also serve as sponsors at Le Mans..

ALETTA: Right. But we focused on the Healey group and a solution was found.

PETER: We were at the Healey meeting in Heidelberg the next year. The USA and Canada came after that.

And how did it evolve?

PETER: After five genuinely successful years with Healey, we wanted to expand our sponsoring. In the course of our search, we found the Peking to Paris classic car rally, which is staged every three years... (1)

ALETTA: ... and for which we served as sponsor and official timekeeper.

What benefits ensued for the Frédérique Constant watch brand?

ALETTA: First, there's the prestige, and then there's the Vintage Racing collection, which was derived from this challenging event.

PETER: When we showed the watches in Basel, a lawyer from Chopard very pointedly called our attention to the already existent Classic Racing collection. For the sake of peaceful coexistence, we renamed our collection "Vintage Rally."

ALETTA: In the future, this name will not only cover our commitment to Healey, but also every other type of rally that we support.

Since 2011 that included La Carrera Panamericana.

PETER: Our Mexican distributor suggested a prestigious rally in his country. I contacted the responsible individuals and we reached an agreement.

In what form?

ALETTA: We became the sponsor and we manufactured a special watch that we could sell for two years. (2)

VINTAGE RALLY
Participants La Carrera Panamericana

Do you want to expand your sponsorship activities?
PETER: We're considering that option. For example, we signed an agreement with Rallye des Alpes...

ALETTA: ... which was cancelled in 2012, although we had already made a watch for it. Now we're hoping the Rallye des Alpes will take place in 2013.

What else can you report?
PETER: In England, we're cooperating with the Grand Prix, a rally for classic automobiles. We'll serve as sponsor and official timekeeper. .

What will the future bring with Healey?
PETER: Our agreement came to an end in 2012. The situation hasn't yet been clarified. We're holding discussions with the family, but thus far nothing essential has been decided about renewing the agreement. It's quite possible that we won't be allowed to sell any more Healey chronographs after 2013. The situation isn't simple, so we we're also supporting other rally events.

ALETTA: In a nutshell, our focus is shifting away from Healey and toward other rallies for which we'll release watches in limited editions.

Will you continue your involvement with the Peking to Paris rally?
ALETTA: That rally will also take place in 2013 and we're currently in the negotiating phase.

PETER: I should explain here that the first time is always the most productive. We can manufacture at significantly lower costs afterwards.

ALETTA: Sometimes we agree to that, but the pleasure isn't the same.

What about antique aircraft? They would be a very good match for Alpina.
PETER: We've thought about it, but other brands are already strongly represented there. That's why I believe we should concentrate on what we've already achieved...

ALETTA: ... which not only includes rallies and runabouts, but also our manufacture work, a ladies' automatic and our classics. These five areas are really enough for us right now.

VINTAGE RALLY
Participants Peking to Paris (2010)

VINTAGE RALLY
Ceremony Peking to Paris (2010)

A watch manufacture must first dream up the idea for runabouts.
PETER: That had to do with our own boat, a 1980 Boesch Cabriolet. We belong to a yacht club and we were asked if we could sponsor an event on Lake Geneva. Without our sponsorship, they said, it wouldn't be feasible. So we agreed.

ALETTA: It was essentially just a small, regional event and it wasn't very expensive.

PETER: Our boat is too young for genuine old-timer regattas. If you want to participate in one of them, you have to own a runabout that was built before 1960, for example, a Chris-Craft, a Garwood or a Hacker. So we capriciously imagined ourselves aboard a fantasy boat and derived our Runabout collection from that vision.

But that's something truly elite for extremely wealthy people. Can a watch brand succeed there?
PETER: Oh, yes. Surprisingly, our Runabout watches are more successful than our Vintage Rallies.

So this would seem to have a bright future?
ALETTA: Definitely. We're also continuing our work on the Vintage Rally collection. The etuis now also include an additional stopwatch. The addition of an automobile was warmly received, but has since been copied by others. The exclusivity naturally suffers when imitators appear on the scene.

Do you dream up these attractive extras on your own or do you collaborate with a team?
PETER: Both. Sometimes our sales personnel contribute ideas that we consider and ultimately implement. We've also received some valuable suggestions from retail customers.

ALETTA: Our advantage in all this is our office in Hong Kong, where our staff can take care of creating and manufacturing these extras onsite in the Far East. We never buy anything that already exists. Instead, we arrange to have the items produced exclusively for us.

PETER: It may be hard to believe, but those are genuinely serious projects. We have to finance the tools for them. We have to take care of the finish for the products. We strongly prioritize worth and preciousness..

How long does something like that usually take?
PETER: Sometimes creating such a box can take just as long as manufacturing the watch itself. One mustn't underestimate the work involved. (1)

(1)

SPECIAL PACKAGING
Luxury wooden gift boxes for our Runabout and Vintage Rally collection

RUNABOUT
Lake Tahoe Concours d'Elegance 2011 & 2012

(2)

WORLD HEART FEDERATION®

(3)

ONLYWATCH

Frédérique Constant supported the ONLY WATCH charitable auction in Monaco

Is there already a collector's market, for example, for the early Healey collections?
PETER: Definitely. Some of those watches have already been sold at Antiquorum and Sotheby's. Naturally, we're proud of that.

ALETTA: But it's also interesting to note that we sell only ten to twelve watches during the events.

PETER: Afterwards, when the edition of 1888 watches is on the market, we don't necessarily sell every one of them.

In such circumstances, it would seem illogical to immediately manufacture all 1888 pieces.
PETER: That's right. We usually start with a batch of 1000. Whether or not we later manufacture the remaining 888 watches depends on the order situation.

In other words, sometimes you don't produce all of the watches in the originally planned edition.
ALETTA: Yes, and that's good for customers who have acquired the watches.

Why did you choose the number 1888?
PETER: That's very simple. We began with 888 pieces because the Chinese believe that 888 is a lucky number. But that number very quickly turned out to be too few watches. So we simply added a 1 at the far left!

ALETTA: Alpina sometimes produces limited series that run for three years. Each is produced in an edition of 8888 units..

You're successful and you allow others to share in that success...
PETER: ...about which we have something very new to report. We'll support the World Heart Federation (2) during the next three years. The contracts have already been written and we donated the first cheque in (mid) 2012, but the official announcement wasn't made until September.

ALETTA: There'll be a limited watch edition in 2013. Then we'll start a collaborative campaign focusing on women's cardiac insufficiencies. With that campaign, we want to heighten women's awareness of how important it is that they stay healthy, for themselves and also for the sake of their children.

You've also participated in "ONLY WATCH" auction. (3)
PETER: ...and we'll again donate a watch to that good cause in 2012. Incidentally, we're also planning to make something unique for the World Heart Federation.

But those are follow-up projects.
ALETTA: No, we've already contributed to the International Children's Heart Foundation, the American Heart Association and the cardiology department at the children's hospital in Peking.

PETER: But the World Heart Federation is much more broadly positioned. It's represented in over 100 countries. There's a foundation in each nation. It's a worthy cause, so we'll provide watches that can be auctioned onsite to raise money for local projects.

ALETTA: We mustn't forget to mention the most recent Passion Awards, which was conferred on the "Paint a Smile" charitable foundation. For us, it's really very important not merely to earn money, but also to spend part of it to support worthy causes.

PETER: Our commitment to charity began with the Passion Awards. It resulted from our motto, which is "Live

CHECK FOR THE IHCF
Nicole Faria - Brand Ambassador and Miss Earth 2010

Professor Bill Novick – Founder International Children's Heart Foundation

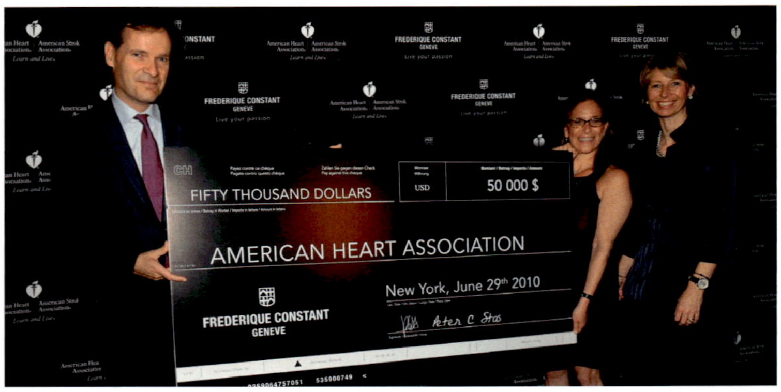

CHECK FOR THE AHA
Helaine Baruch
VP American Heart Association

DONATION BEIJING CHILDREN'S HOSPITAL
Inauguration of Paint a Smile's decoration on the walls of the cardiology department of Beijing Children's Hospital with ShuQi, brand ambassador Frédérique Constant.

Professor Bill Novick
Founder International Children's Heart Foundation

Your Passion." An award for impassioned commitment fits that perfectly.

ALETTA: We published our intentions in the "Herald Tribune" and other newspapers. The jury, which includes Peter and me, chose the winners from among the suggested candidates.

PETER: The first winner was Peter-Frans Pauwels, who invented the TomTom navigational system.

ALETTA: After the award's conferral, someone called our attention to nonprofit and charitable organizations. He said that this was another field in which people are passionately involved. I must admit that we hadn't thought about that at first, although far greater passions are often involved there.

PETER: Dr. Bill Novick is a wonderful example. Alongside his professional career as a physician, he's also passionately involved in the International Children's Heart Foundation.

ALETTA: In our opinion, this professor definitely deserves an award for his dedication. So we're turning in the direction of charity and we're very happy about that. We sometimes need suggestions and impulses because they open our eyes. That's how we learn. After all, so many people on this planet desperately need help.

But you also donate profits from sales of your watches.
PETER: Motivated by our experiences with the Passion Award, fifty Swiss francs of the revenue earned from the sale of each Double Heartbeat watch are donated to charity. Dr. Novick received one check and a second one was donated to "Paint a Smile."

ALETTA: One day we noticed that many famous people are also active philanthropists. When we developed the Chinese market, we became aware that we urgently needed a brand ambassador. That's not absolutely necessary here in Europe, but it's tremendously helpful in the Far East.

PETER: We contacted Miss ShuQi and an agreement ensued. She's a very well known and extremely popular actress in China and she was personally present when we gave our donation check to the "Paint a Smile" foundation.

ALETTA: She joined us on a visit to young patients in the cardiology department of the children's hospital in Beijing. We're very happy with Miss ShuQi and she seems equally happy with us because she recently renewed her ambassadorial contract. I suppose it's important to Miss ShuQi because alongside her luxurious world, our support enables her to become involved in another, less glamorous side of human life.

PETER: It's simply different when you can personally experience things and see how much joy you can bring to people who are less fortunate.

Would you care to sum things up?
ALETTA: Collaboration helps both sides. That much is clear to me. I almost forgot to mention Nicole Faria, Miss Earth and Miss India. She also assists us as brand ambassador and accompanies our campaigns to help children with cardiac insufficiencies.

CHARITY
Miss ShuQi in Beijing's Children Hospital (2010) (t), Giving gifts to children (lb), Press Conference (mb), FC-310LHB2P4 (rb)

中国
中央电视台

FREDERIQUE CONSTANT
GENEVE

FREDERIQUE CONSTANT
Awards
for Charity

PASSION AWARDS

If Frédérique Constant stands for high quality manufactured Swiss watches, it is first and foremost the story of a passion. Indeed, it all started with our passion, back in the 1980's, as we were drawing watches in our spare time, after a long day's of work.

Many elements inspire success in business. Product, plan and ambition play their part.

But the greatest predictor of global accomplishment is the hardest to acquire. That factor is passion. Passion for your product, passion for your mission, passion for business itself.

We believe that the overriding factor for our success has been passion. The passion which turned us from lovers of fine watches to global producers. Over the years, we realized that a company, organization or person is rarely praised for their passion. In fact, passion was vastly underestimated. This is why our company slogan has always been, "Live your Passion," and why we created the Frédérique Constant Passion Awards. We want to recognise businesses, organisations and persons fired by passion and proving the entrepreneurial success which that elusive quality brings.

PASSION AWARDS
Passion Awards for Charity
in Cipriani Wall Street 2010

Learn and Live
Live your passion
FREDERIQUE CONSTANT
GENEVE
Live your passion
American Heart Association
American Stroke Association
Learn and Live

FREDERIQUE CONSTANT
GENEVE
Live your passion
FREDERIQUE CONSTANT
GENEVE
Live your passion
FREDERIQUE CONSTANT
GENEVE
Live your passion

CANDICE ACCOLA
Actress Candice Accola at the Passion Awards for Charity 2010

Winners of the first Frédérique Constant Passion Awards were Peter-Frans Pauwels of TomTom and Dr. William Novick of the International Children's Heart Foundation. The prize was a business case study by Prof. Dominique Turpin of IMD in Lausanne on a strategy issue of choice, an invitation to our Award ceremony and concert in Geneva and a gold Frédérique Constant watch from our Heart Beat Manufacture collection. The Financial Times was the official media partner. The award ceremony and concert was organized in partnership with Fortis Banque (Suisse) SA and ABN AMRO, who were also sponsors of the non-profit prize.

OBJECTIVES OF THE FRÉDÉRIQUE CONSTANT PASSION AWARDS ARE:

• To highlight the true passion that is typically the main driver of a successful entrepreneurial business or organization

• To recognize annually outstanding entrepreneurs who successfully develop from start-up to next phase fast-growing business or organization

• To identify best practices which may be useful to other entrepreneurial businesses or organizations

CRITERIA

Candidates and their organizations are evaluated on the following criteria:

• Success is the result of true passion and drive of entrepreneur(s) who founded and are still managing the company or organization.

• The candidate has a strong ability to communicate his or her passion

• Has achieved solid results and continues to grow well beyond the average.

• Has a strong ability to innovate and sustain their company or organization business model.

• Is an international business or organization.

• Has established an effective governance system and demonstrates responsible ownership.

WILLIAM NOVICK
Founder of the International Children's Heart Foundation and winner of the Frédérique Constant Passion Awards 2007

HANS-PETER BOLLINGER
Founder of Microstep and finalist of the Frédérique Constant Passion Awards 2007

LADY DOUBLE HEART BEAT
FC-310CDHB2PD4

FORTIS
EDERIQUE CONSTANT
GENEVE
Live your passion
Banking

FREDERIQUE CONSTANT
GENEVE
Live your passion
American Heart Association
American Stroke Association
Learn and Live

PROCESS

PIERRE MARCOLINI
Founder of Marcolini Chocolatier and winner of the Frédérique Constant Passion Awards 2008

LAURA COTTON
Founder of the Paint-a-Smile Foundation and winner of the Frédérique Constant Passion Awards 2009

HELAINE BARUCH
VP American Heart Association receiving Frédérique Constant Passion Awards 2010 donation

In order to find the most meaningful candidates, the following submission and selection process was established:

• A call for submission of candidates was published.

• The Jury for the Frédérique Constant Passion Award evaluated submissions. An in-depth assessment matches each candidate with the six qualifying criteria.

• The ten highest qualifying candidates were announced, and subsequent additional evaluation by the jury determined the final winner.

• The qualifying candidates were contacted for interviews and complementary information which was not publicly available.

• The qualifying candidates were invited to visit the Frédérique Constant factory in Geneva and attend the Award Winning Ceremony.

Winners of subsequent Passion Awards were Pierre Marcolini, the founder and CEO of Pierre Marcolini Chocolatier, Laura Cotton, the founder of Paint a Smile, a foundation dedicated to the improvement of hospital environments, and Helaine Baruch of the American Heart Association.

Thanks to the Passion Awards, we have met many very passionate people with beautiful stories. These people are some of the most passionate people and organizations in the world, and meeting them was an unforgettable experience. That is why we plan to continue with the Passion Awards. We hope to meet more people who are driven by passion and that more companies, organizations and people realize success through passion. Throughout our lives, we have learned one thing, it doesn't matter if you are a heart surgeon or a chocolatier or a watch manufacturer. What matters is if you are passionate about what you are doing.

FREDERIQUE CONSTANT
GENEVE

EVA LONGORIA
Par Coeur Gala 2010 – Paris

ALETTA AND PETER STAS TALK ABOUT

FAMILY, CHILDREN AND FRIENDS

Is it more difficult on the market for family businesses than for big corporations?

PETER: Basically, yes. We primarily work with agents, some of whom are better than others. Our distributors generally work very hard on our brand's behalf, but we still have to conduct plenty of training sessions to highlight the achievements and remarkable values of Frédérique Constant.

ALETTA: The alternative would be to set up our own subsidiaries, but we don't want to go that route because it would be too costly and would entail a much greater investment of managerial effort. We already have three subsidiaries and I always emphasize that there won't be any more in the future.

INÈS SASTRE,
Frédérique Constant
Charity Ambassador

So big corporate groups have an advantage?
PETER: Yes, definitely. Groups own subsidiaries which carry many brands and can therefore work much more efficiently. Everything is optimally organized.

ALETTA: The second advantage involves collaborative purchases of advertising. Because of their large size, they know exactly what they're buying. As a comparatively small business, we're weaker there. First, we cannot buy so much. And second, we don't have much experience with media planning. Unfortunately, a lot still occurs by trial and error.

PETER: Finally, big corporations can exert group pressure. If you're a jeweler and you want to carry brand A, you're compelled to also carry watches from brand B. This coupling can be very aggressive. They sometimes go so far as to insist that jewelers either remove Frédérique Constant from their assortment of merchandise or risk losing the privilege of carrying watches made by successful brands that belong to big corporate groups.

ALETTA: When a jeweler is put under such strong pressure, he may regretfully have no choice but to turn us away because he simply cannot afford to lose one or another of the big brands. That's the situation in the business these days.

PETER: So we rely on smaller family-owned jewelers who also order from corporation-independent suppliers and can thus cultivate a healthy mix of brands.

An alternative would be to operate your own boutiques.
PETER: We've opened boutiques in Seoul, as well as in China, Hong Kong and Saudi Arabia. Our agents operate them. We also have at least 150 shop-in-shop points of sale, where we're much more successful than in our own boutiques.

Why?
ALETTA: It's not easy to lure customers into a single-brand boutique. We're simply not yet known well enough. There's also the problem of costly rents and high expenses. With our moderate price level, proprietors must sell very large numbers of individual watches to make a profit. The situation is different at a shop-in-shop because customers are already in the store. They stroll by, notice us, feel interested in our products, receive well-targeted advice ... and buy our watches!

Sooner or later, every purchase is followed by customer service.
PETER: Frédérique Constant guarantees all its products for a period of two years. If there should be any sort of issue with one our watches, fifty service centers around the globe are available to help our customers. All centers are networked with our headquarters. If they don't have the parts they need, we send the components without delay. We also offer a repair service in Geneva for more complex watches such as tourbillons. Each watch receives a tracking number so the customer can follow the progress of his watch's repairs via the internet. (1)

How do you handle claims for compensation of the sort that occasionally arise with watches sold under a two-year guarantee?
ALETTA: Yes, there unfortunately are some customers who hurl a watch to the ground or otherwise mistreat it. They then claim that the product was at faulty and they insist that the damage is covered under the guarantee. In such instances, we closely scrutinize the watch in our factory. After ten years of experience, we can correctly determine the cause of the damage.

PETER: But ultimately we prefer not to fight about it. We would instead find a generous way to settle the dispute.

Have there been any really extreme cases?
ALETTA: About once a year, a truly aggressive customer will contact us. We usually opt to return the purchase price. We have over 100,000 satisfied customers each year, so a single angry one doesn't tip the scales very much.

ALETTA: Less than one percent of our merchandise is returned to our factory. Our service centers also perform repair work, but returns are very minimal. The errors in our manufacture are manageable.

Why is that?
PETER: Frédérique Constant continues to produce a large number of quartz watches and timepieces with comparatively simple self-winding mechanical movements, both of which are inherently less vulnerable to errors. The quality of each watch is meticulously controlled before it leaves the factory. And we seldom have problems with watertightness.

ALETTA: Watches with Frédérique Constant's insignia are watertight, but we deliver all our watches with leather wristbands, so our customers usually take them off before they go for a swim. That's a big advantage.

PETER: We've observed that Frédérique Constant's customers almost always treat their watches with care. Problems with watertightness are quite evident in general. Sport watch brands can tell you all about it. We're seeing it with Alpina.

(1)

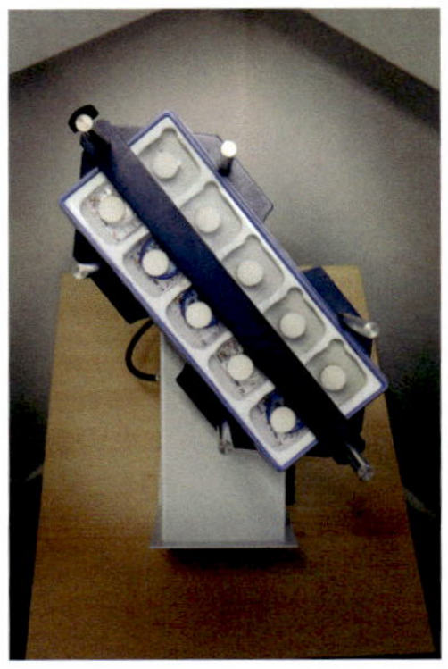

Quality Control in the Rotating tester

MONO BRAND BOUTIQUE
Opening ceremony of the Frédérique Constant Boutique in Taiyuan
2011

FREDERIQUE CONSTANT
GENEVE
康斯登
FREDERIQUE CONSTANT
GENEVE

生意兴隆
财源广进

PIETER-JAN & ELINE
Next generation?

Do you have friends here?
PETER: Yes, quite a few, including a few Geneva natives. It's always possible to make friends if you're adaptable and you make the effort.

Is there a Dutch community in Geneva?
ALETTA: Yes, quite a large one, in fact. But we don't participate in it much.

Why not?
PETER: The atmosphere here is different than in Hong Kong's Dutch community. That bothers me a little, although I can't really describe it. We have Dutch friends here, along with our German and Swiss friends.

How do you find the public administration and the governmental authorities here in Geneva?
ALETTA: They've undergone big positive changes in recent years. When we first arrived, we had to show them our business plan for the coming five years – and we weren't sure they'd accept it.

PETER: Wealthy people have it easier because they bring money and they pay money. Our situation was different. We basically didn't possess anything at that time, but we hired a lawyer with relevant experience and he took care of it for us.

You own a family business, so you must also think about the next generation, especially because you're not planning to sell Frédérique Constant. You have two children a boy and a girl. Do they like watches?
PETER: Our children are still young. Basically they have fun with watches. Our 16-year-old son is more interested in science and our 12-year-old daughter would be more likely to take an interest in watches. But at such young ages, it's too early to begin thinking about that.

Are you bringing up your children so they'll one day take over your business and continue it?
ALETTA: We're taking a rather passive approach with their upbringing, as far as someday taking over our company is concerned. At home, they already have to listen to so many stories from the daily life of a watchmaking manufacture. There's no way to avoid that, but we aren't actively bringing them up in this direction.

PETER: When the time comes for them to decide, we want our children to be totally free to choose what they want to do and what they would prefer not to do.

Are they interested in knowing what's happening at Frédérique Constant and how the business is developing?

ALETTA: Basically, yes, but they don't say." Now I'm going to be the boss." They're still too young for that. But our son was here many times during summer vacations and he helped at Alpina. He worked on the apps. In this way, he gets to know that an entrepreneur has to work very hard and with unflagging concentration to make sure that nothing drops into his lap by mistake! As the saying goes, "Nothing's for nothing."(1)

(1)

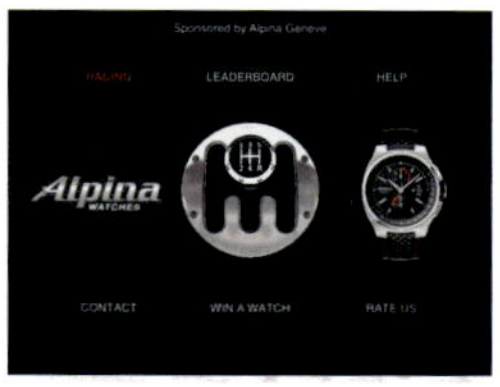

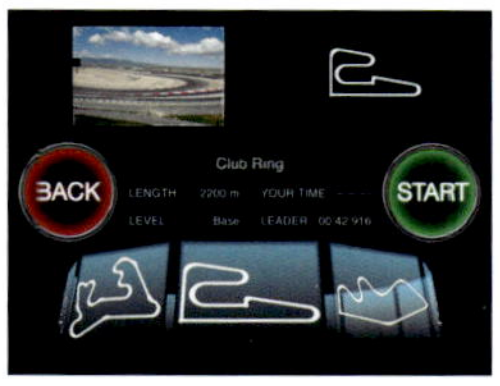

Screens from the Alpina iPad Application co-developed by Pieter-Jan Stas

As a mother, would you be pleased if your children were attracted to the business?

ALETTA: Yes, of course, how could I not be pleased by that? It would be wonderful. If it works well, it's fantastic.

What is it like for a married couple to share their business life and to work so closely together?

ALETTA: Now and then I hear people say it must be problematic. I admit it can be difficult at times. But when it works smoothly, there's definitely nothing better.

PETER: We share the same passions and we also share the occasional worries. That makes things much easier.

ALETTA: I see things similarly with regard to the children. If they choose to join the business and everything works smoothly, it would be marvelous. But suppose they graduate from university and take jobs with companies located somewhere halfway around the globe. Then we would see them only rarely. It would be much nicer to have our children working under our own roof.

Do your opinions about Frédérique Constant sometimes differ drastically from your husband's views?

ALETTA: That naturally happens now and then. But I know how to convince him that my opinion is right! Seriously though, our views seldom diverge extremely.

PETER: Differences of opinion are most likely to occur when we didn't have the time, or didn't take the time, to sit down together and discuss the problem in depth. We each have our own clearly defined fields of activity in which we think and act in a results-oriented manner. Often the disagreement is really only about which path is the best route to our shared goal.

ALETTA: The form of the dialogue is always important because we both already know the contents. How should I phrase what I want to express? There's a certain difference here depending on whether you're talking with a colleague or with your spouse. Even if we didn't work together, the ways and means of our communication would be decisive for our mutual understanding. I believe that the right way of interacting can solve at least half of all problems.

If you had three wishes for your future and for the future of Frédérique Constant, what would they be?

PETER and ALETTA:

1. That we continue to grow;

2. That there will always be a place for independent, family-run watch manufactures because their existence encourages creativity in general;

3. Finally, that all people on our planet can enjoy good health. Through our charitable activities, we see time and again that people, especially children, suffer from cardiac insufficiencies. Within our modest means, we want to continue to do all we can to combat that problem.

SHUQI
Actress and Frédérique Constant brand ambassador since 2008

FEMININE LUXURY FOR THE WRIST

"LADIES' AUTOMATIC COLLECTION"

Some say women aren't interested in mechanical watches, but nothing could be further from the truth! The world has changed and watch collections such as Frédérique Constant's "Double Heart Beat" spark enthusiasm among members of the fairer sex.

Many women believe that a wristwatch should be pretty to look at, able to survive occasional harsh treatment, and practical to operate. But let's examine the facts of the matter more closely. Passion and interest in a high-quality mechanical wristwatch don't materialize out of thin air, but grow from a maturation process that usually occurs unseen. The catalyst is often a mutual passion cultivated by a couple. He discovers his love for mechanical watches: she catches the bug from him: her interest grows and soon her passion's intensity rivals his. When a father, a mother or both parents

FREDERIQUE CONSTANT

CONSTANT

collect mechanical watches, it's only natural that their daughter will become fascinated too. Seeing a beautiful timepiece on a loved one's wrist awakens a woman's desire to own a fine watch of her own. A dynamic and independent businesswoman may wish to impress her male colleagues. And a mechanical wristwatch is a gift that aptly expresses love. As time passes, its recipient discovers and surrenders to the fascination of mechanisms. Ideally, the manufacturer's advertising sparks interest and the specialized trade further fans the flames. Many roads lead to the same destination. Awareness must first be raised about the mechanisms under the dial. Only afterwards does a person comprehend the fine differences between a quartz watch and its mechanical counterpart. Once this consciousness has been raised, he or she soon loses interest in quartz timepieces. It's not without good reason that aficionados regard quartz watches as lifeless, inert matter.

LADIES AUTOMATIC
Heart Beat iconic and feminine timepieces - Made by women... for women

Frédérique Constant discovered the fascination of mechanical timekeeping at an early stage. The dial was pierced at precisely the location where the aperture would reveal the escapement and prove that this watch has a beating "heart." This innovation came in 1994, soon after the end of the dull era of the quartz watch, which nearly caused mechanical watches to become extinct. The keyword here is "heart" because a mechanical movement really does have a heart--namely an escapement. Its parts are the pallets, the escape-wheel and the balance, which is composed of a balance-wheel and a hairspring. When the balance tirelessly swings to and fro in uniform semi-oscillations, the balance-wheel alternately rotates through nearly a half-turn in the clockwise direction, immediately followed by an

FREDERIQUE CONSTANT
26 TWENTY-SIX JEWELS SWISS MADE

identical but counterclockwise near half-turn. Meanwhile, the paper-thin hairspring contracts and expands over and over again. These uniform concentric motions can be compared to the beats of a healthy heart, which indefatigably repeats unvarying pumping motions that cause blood to circulate through our body. The choice of the name "Heart Beat" for this new type of watch was thus not coincidental. The aperture in the dial invites connoisseurs to peer directly into the "heart" of the mechanical watch, where they can admire the working of its balance and view the pump-like motions of the spirally coiled hairspring through a loupe. Unlike a prosaic quartz watch, this type of timepiece is vividly alive. Nowadays, concepts such as "Heart Beat," "Double Heart Beat," "Love Heart Beat" and "Amour Heart Beat" designed by ShuQi have evolved into permanent features in the identity of Frédérique Constant. And the people in Geneva, where Frédérique Constant is headquartered, are justifiably proud of these achievements. Imitation is the sincerest form of flattery and good ideas typically attract imitators. The aperture in the dial wasn't protected by a patent, so competitors soon began cutting openings into their dials too. Be that as it may, there's only one original. And it's made solely by Frédérique Constant!

AUTOMATIC
Caseback view of the Ladies Automatic collection

FREDERIQUE CONSTANT
GENEVE
SWISS MADE

Now the moment has arrived to switch on the spotlights, draw back the curtain and prepare our stage to welcome the glamorous world of ladies' watches from Frédérique Constant. A bevy of beauties vie for the honor of gracing m'lady's wrist. Three of these models are named "Lady Double Heart Beat," "Lady Love Heart Beat" and "Amour Heart Beat designed by ShuQi." There's also a fourth version with a closed dial called the "Lady Automatic." It's difficult to choose one's personal favorite from these beautiful candidates at first glance. Let's see if closer scrutiny can alter this situation.

The "Lady Double Heart Beat" models have earned a special status in the collection of ladies' timepieces. Cult status has long since been achieved by the two interlaced hearts that form an opening directly above the movement's escapement at "12 o'clock." This model symbolically unites the two hearts of a loving couple and simultaneously offers an unobstructed view of the escapement so the "heart" of the automatic movement can be admired. The assortment currently includes four versions: a 34 mm variant with a steel case and a grey mother-of-pearl dial with eight diamonds to serve as its hour indices; the same model with a gold-plated steel case and a vanilla-colored mother-of-pearl dial, likewise adorned with eight diamond indices; a richly decorated model in a gold-plated steel case with 48 diamonds on the bezel and 53 additional diamonds on the inner part of the chocolate-colored mother-of-pearl dial, which relies on eight additional diamonds to mark the hours; finally, the selection concludes with this last-mentioned model, but without the 53 diamonds in the center of its mother-of-pearl dial. An appropriately colored cabochon gemstone decorates each model's winding-stem.

BLACK BEAUTY
Double Heart Beat "Black Beauty"
FC-310BDHB2PD6 (2012)

FREDERIQUE CONSTANT
GENEVE
SWISS MADE

The "Lady Love Heart Beat" models let no doubts arise about the identity of the woman to whom their dials' distinctive message is addressed. The word "love" is skillfully and stylishly staged here. Yes, you guessed correctly. The "O" in "love" forms the famous opening in the dial, through which the heart of the automatic movement can be observed. A more intimate and simultaneously ticking declaration of love would be difficult to imagine. Here again, the selection includes four models, each with a 34mm-diameter gold-plated steel case. One variant has a vanilla-colored dial on which 64 diamonds spell out the word "love." A second version features an identical dial surrounded by a bezel adorned with 48 diamonds. If desired, each of these models can be combined with a gold-plated steel bracelet. Finally, the selection is rounded out by a variant with a chocolate-colored dial embellished with six diamond indices, a bezel with 48 diamonds, and the word "love" formed by 50 diamonds on the dial.

LOVE HEART BEAT
FC-310LHB2PD6 (2007)

FREDERIQUE CONSTANT
GENEVE
VI
SWISS
MADE

Paris is the city of love and French is its language.

Paris is the city of love and French is its language. Who among us hasn't whispered "mon amour" to his beloved? Whether the word is "amour" or "love," the emotion is the same. And it's lauded here by the Chinese actress ShuQi, who created the "Amour Heart Beat designed by ShuQi" collection. Like their counterparts on the dials of the Lady Love Heart Beat collection, numerous diamonds spell out the word "amour" on each dial. A total of 85 gems beautify the case and dial. This 34mm-diameter model is available in three variants, each of which is manufactured in a limited series of 888 timepieces and all of which reveal the ticking hearts of their automatic movements. One version combines a steel case with a pale grey mother of pearl dial, another weds a gold-plated case to a vanilla colored mother of pearl dial, and the third boasts a chocolate-colored mother of pearl dial.

AMOUR HEART BEAT
Limited Edition designed by ShuQi

FREDERIQUE CONSTANT
GENEVE
Automatic
24
SWISS MADE
FREDERIQUE CONSTANT
GENEVE
Automatic
25
SWISS MADE

An aficionada who prefers a more traditional look can choose the "Lady Automatic," which has no aperture in its dial, but instead provides the convenience of a date display. Three classical variants of this self-winding wristwatch are available. The first is a steel version with a lavishly diamond-decorated bezel and a mother of pearl dial bearing ten diamonds which serve as hour indices and contribute toward a grand total of 58 diamonds. The second version has a gold-plated case. The wristbands too are freely combinable whether a steel bracelet or a satin strap, both are possible. The broad spectrum is rounded out by the third option, which features a chocolate-colored dial on which ten diamonds form the hour indices.

May we offer even more? All models in the "Ladies Automatic Collection" are equipped with the so-called "Easy Release System," a user-friendly mechanism that allows the wearer to readily remove one wristband and replace it with another. Regardless of whether she's at home or away from home, a few simple motions suffice to change the wristband. Several bands in various materials and colors are available to complement each model, so each of these watches can be effortlessly modified to match madam's momentary wardrobe. If she's in the mood, she can also choose an especially trendy strap to create a colorful accent. The gift boxes in which the watches in the "Ladies Automatic Collection" are delivered, double as gorgeous travel etuis that not only provide space to store the watch and several wristbands, but also offer room for their owner's personal jewelry. Here are practicality and prettiness, all rolled into one.

The rivalry for the coveted place on a lady's wrist ends here. Now a woman is more likely to ask a different question. With such an enormous selection of ticking beauties to choose from on which lucky watch should I first bestow my smile? Indecision can be painful, but that's the exquisite agony of choice!

LADIES AUTOMATIC
FC-310CHD2P4 and
FC-310WHD2PD4
2008

FREDERIQUE CONSTANT
GENEVE
SWISS MADE

RUNABOUT

A VOYAGE IN TIME

First built in the United States in the early 1920s, the speedy and luxurious powerboat that came to be known as the runabout carries the genes of the small horse-drawn carriage that was its distant European ancestor, before the latter crossed the Atlantic with the pioneers of the American West.

The runabout has always been a restless explorer, a lover of freedom and the great outdoors. Its earliest manifestation, drawn by a single horse, was game for a brisk excursion in the 19th century English country-side. Later, in an epidemic of Anglomania, some French inventors of the automobile gave the name to a series of small vehicles that, though backfiring and unpredictable, nevertheless offered freedom at last from the constraints of horse-drawn transport. But while the European elite remained happy indulging in a spot of Sunday mechanics, the American pioneers in their vast roadless expanses retained their affection for the horse and cart for a long time: for going to the local shops at a gallop or visiting friends. The method of transport blended utility with pleasure. Or, as these tireless colonisers were wont to say, "running about and getting things done."

World War I would change the face of the world- the rural as much as the industrial world, the east as much as the west. Some historians would go so far as to say that the 20th century only began in 1914. In victorious America four years after the start of World War I, the country's new prosperity was founded on three pillars: the market, the motor engine, and oil, to the point where the legend printed on the dollar might well have been changed to "In Oil We Trust."

RUNABOUT VENICE
"Venice Limited edition"
FC-392RV6B6 with special engraving on the caseback
(2012)
Limited edition of 1888 pieces

page 154-155
RUNABOUT
Historical photos of runabouts in Lake Como, Italy in the 1930's. Count Theo Rossi di Montelera winner of the President's Cup in Washington in 1938 (mb). Gar Wood triple cockpit (rb).

AU GRAND PASSAGE
Firestone
LAURENS
ZURICH
VITA

58

Scafo
San Marco
57
MaMi

57
SanMarco

57
MaMi

TOWARDS ACCESSIBLE LUXURY, FROM THE USEFUL TO THE PURELY ENJOYABLE

From that time, the New World only wanted to look to the future, and weekend leisure became the legitimate reward for a well-spent working week. Motorised sports, on water as well as land, experienced unprecedented growth. Assisted by Christopher Smith, a man named Garfield Wood, a billionaire inventor, installed light, powerful aircraft engines in wooden boat hulls that were long and almost flat. The resulting rustic-looking craft skimmed across the surface of the water at an almost unimpeded pace, which was soon to surpass the record-breaking speed of more than sixty miles per hour! Chris Smith was not a man to let an ingenious idea pass him idly by. While driving a fast boat can be insanely exciting, it can also be useful, as long as it is comfortable and well-equipped. There was no shortage of navigable lakes and rivers in his country, and what's more, many of those who had achieved success also opted to live near water and already owned at least one car. And so in about 1920, the modern motorboat was born that could be driven as easily as a car. It would be called the "runabout," and used simply for making a splash, cutting a dash, for visiting friends, camping, fishing, or, as the grandparents would have it, "running about and getting things done!."

RUNABOUT
Photoshoot new Frédérique Constant Runabout campaign

APACHE II

THE RUNABOUT LIFESTYLE

The 1920s and 1930s saw the new man in flying form, liberated and mobile. Not for him the old pocket watch. He wore time on his wrist, like an aviator, the real hero of the day. Distant voyages beckoned, for the planet had not yet delivered up all its secrets and seemed to be without limits. Runabouts were becoming more sophisticated and powerful, adopting the lines that were in vogue at the time. They were modeled based on concepts from the great creations of Art Deco or they borrowed the streamlined silhouettes of futuristic design. Speed, that barometer of freedom, was evoked in new forms that sought to give a meaningful dimension to the most everyday functions. In a runabout, settled into deep leather seats and protected by the windshield, the passage of time was recorded by dashboard clocks framed in round, square or rectangular chrome. But these times also belonged to writers and artists who travelled the world. Far from being merely only an amphibious New York-style limousine for the likes of Scott Fitzgerald's Great Gatsby, the runabout was also a faithful companion when conducting globetrotters to the seaplane awaiting them in the bay of Hong Kong or Rio. The same varnished mahogany glinting in the moonlight could bring you alongside the impressive yacht resting serenely at anchor, to which you were an invited guest. Remember the time Paul Morand, that quintessential man of the world, was following a detour through the Singapore Yacht Club; a few minutes later, he was drinking a cocktail in the company of Somerset Maugham at the bar of the Raffles Hotel. Meanwhile in Cannes, Jacques-Henri Lartigue, while aboard a runabout, captured perfectly in black and white a moment of pure felicity when he set his lens on a hull gliding past the Carlton. For in Europe, the motorboat, a mark of elegance that was frequently an American import or clone, created by Chris-Craft, Gar Wood, Boesch or Corsier Port, remained the preserve of a wealthy elite on the lakes of Zurich, Como or Geneva.

RUNABOUT
Lake Tahoe event (t), Runabout on Lake of Geneva (m), Poster Helice Nautique (lb), Frédérique Constant Main Sponsor Helice Nautique at the Societe Nautique Geneve (rb)

THE TOMBOY COMES OF AGE

From the end of the 1920s, with her short hair hidden beneath a cloche hat that only partially eclipsed the brightness of her clear eyes, Delphine Dodge, daughter of the founder of the famous automobile brand, was one of the first celebrity figures to take to the cockpit. She won regattas in Havana and Palm Beach, navigating stylish runabouts built by her brother Horace under the tutelage of John Hacker, a genius at the drawing board. Her friend, Maud Hughes, and others did likewise on the Potomac, the St. Lawrence and the French Riviera. But the laurels for headline-grabbing exploits aboard a powerboat must go to the tempestuous Briton, Betty Carstairs. Armed with considerable courage and a solid fortune inherited from her oil mogul grandfather, she was determined to beat the top American champions of the day. Before retiring to her private island in the Bahamas, she maintained for many years on Lake Geneva a fabulous 9-metre mahogany motor boat. The runabout was powered by a Napier Lion aircraft engine, which she was testing for a record-breaking transatlantic endeavour.

PASSION FOR DETAILS

(1)

Historical Runabout moo-ing

(2)

Historical Runabouts in _ake Como, Italy.

ITALY LEADS THE WAY TO SAINT TROPEZ

Along with the end of World War II and the prosperity of the post-war boom came the rise of the seaside culture. The runabout was at the heart of all this beach resort activity, but it had changed. The new model was more likely to have come, not from America, but rather from Sarnico, Italy, where Carlo Riva was now setting the pace. With Riva's high-end models like the Aquarama, the 1960's powerboat took off as the ultimate pleasure craft, laden down with leisure time pals and bikini-clad girls stretched out on the rear padded sunbathing bay. More than ever, the runabout symbolised both personal success and a mastery of movement. As the cult of the individual came to dominate that of the community, the runabout facilitated escape from the crowd, from the congested shore, as well symbolizing differentiation from the masses. The playboys of Saint Tropez -- people like Roger Vadim and Gunther Sachs -- were partying, and their gleaming Rivas were everpresent with them in the dream-like photos taken by the popular press every summer. Then, following a brief eclipse in the 1980s and 1990s, the classic varnished wooden runabout made a comeback, first in America among passionate collectors, and then across the Atlantic, through the restoration of boats from a time we had thought long gone. It is not just the vintage version, however that is enjoying a second wind. A few boatyards, including Switzerland's Boesch, continue to produce new varnished and modern wooden runabouts, while others launch new models in ultra-contemporary materials. Riva, in keeping with its long tradition, has since 2000 been producing an updated version of the new era's runabout in the form of the Aquariva. (1)

ART AND THE RUNABOUT IN THE SPIRIT OF THE TIMES...

Some artists and designers, seized by an inspiration that has little to do with nostalgia, have adopted what may have been perceived as an aging relic to inject it with what looks like the makings of a brightly promising future. Xavier Veilhan, for instance, who was the guest of Versailles in 2009 as the successor to Jeff Koons, in 2010 created the first monochrome blue RAL 5015 speedboat named after Yves Klein, in collaboration with the Austrian boatyard, Frauscher, in 2012. A second limited edition in a variant blue is in production. Meanwhile, Riva has unveiled at the prestigious fine art Gagosian Gallery on New York's Madison Avenue, among Picasso, Cy Twombly and Robert Ryman paintings, Australian designer Marc Newson's minimalist interpretation of the Aquariva runabout, which will only be produced one unit at a time, in the same manner as a work of art. Moreover, as recent auctions have demonstrated, art and the runabout are not exactly unrelated in concept. When Sotheby's was engaged to dispose of Gunther Sachs's art collection in May 2012, the Super Ariston owned by the king of long nights in St. Moritz achieved almost €500,000, five times its estimate. In classic or ultra-contemporary form, the runabout has not yet lost its ability to defy time. (2)

RUNABOUT MOONPHASE
FC-360RM6B4 in its luxurious wooden gift box with the boat miniature (2011)

FREDERIQUE CONSTANT RUNABOUT

(1)

Pictures from the Frédérique Constant Runabout Challenge in Venice (2012)

Since 1998, we have been owners of a Boesch 650 Cabriolet Runabout, and we have also been members of the Societé Nautique, de Genève (SNG), the Geneva yacht club. The organizer of the Hélice Nautique, an event honouring runabouts and organized by the SNG, asked Frédérique Constant to became the Official Timekeeper of this event. The Hélice Nautique brings together many of the remaining Runabout yachts on Lake Geneva, yachts that today, as in their heyday, symbolise elegance, sophistication and ageless design. Hence, 2009 saw the introduction of the limited edition "Runabout" watch series from Frédérique Constant, designed to celebrate the brand's involvement in the annual Hélice Classique. An elegant duo of automatic chronographs was created with a design perfectly adapted to the fascinating and nostalgic, teak-wooden luxurious motor yachts of the 1920's. In solid steel or plated with rose gold, the rounded case with a diameter of 43 mm is hand-polished and exposes a finely decorated dial with a "Pavé de Paris" guilloché decoration. The applied and polished indexes, following the material used for the case, guarantee clear readability. As a delicate hint to the maritime theme, a boat's propeller is engraved on the case back and a "porthole" offers a glimpse of the movement. To emphasize on the special value of this timepiece, it is delivered in a gift box that contains a miniature duplicate of a legendary "Runabout" yacht. (1)

In later years, Frédérique Constant also became the Official Sponsor of the 39th and 40th annual Lake Tahoe Concours d'Elegance Wooden Boat Show, held on the north shore of Lake Tahoe in Carnelian Bay, California, USA. New versions of the bestselling collection were introduced, including the Runabout Automatic, the Runabout Power Reserve and the Runabout Moonphase. Each Frédérique Constant Runabout is limited to 1888 pieces and delivered with a miniature runabout model.

RUNABOUT VENICE
FC-392RV6B6 and FC-303RV6B6
Runabout "Venice Limited edition"
Limited edition of 1888 pieces
for each model (2012)

FREDERIQU
NSTANT RUNABOUT
No.0001
LIMITED EDITION 1888
FC-303RV6B6
10 ATM
SWISS MADE

FREDERIQUE CONSTANT
GENEVE
FREDERIQUE CONSTANT
GENEVE
SWISS
MADE

LIVE YOUR PASSION

DRIVE A CLASSIC MOTORCAR!

The world of mechanical watches and the cosmos of historical automobiles are closely interwoven. Their common denominator is always identical -- passion!

VINTAGE RALLY
Vintage Rally La Carrera Panamericana limited edition FC-435S6B6 with its Stoptimer (2012)

What makes a classic motorcar so irresistibly fascinating?

What makes a classic motorcar so irresistibly fascinating? We forgive nearly all its blemishes and we good-naturedly put up with its idiosyncrasies, even though we would never dream of being so lenient with a modern car. Why do we voluntarily climb aboard a historic vehicle and pilot it through traffic, even though it lacks every comfort, convenience and safety feature of a contemporary automobile? Why do we forget all about icy winds and inclement weather while we blithely cruise across the landscape with our classic car's top down? The reasons are irrational. It's a passion, and you either have it or you don't. A wristwatch can also be worn merely to tell the time, just as a car can be used simply to convey us from point A to point B. These are both possibilities. Alternatively, rather than simply wearing a watch on one's wrist, one can be actively and profoundly interested in it and especially its inner life. Analogously, one can enthusiastically derive pleasure from one's automobile and from driving it, sometimes passionately and sometimes speedily, in harmony with the doctrine that the path is the goal. These are the other options. Passion frequently plays a major role here, so it cannot be merely coincidental that the ardor for automobiles and the fervor for mechanical wristwatches meet each other eye to eye. Whether it's a contemporary car or a historical automobile, the two worlds simply fit together perfectly.

An Austin Healey motorcar is one such mythical and simultaneously irrational vehicle that's guaranteed to stop us in our tracks whenever we see one. We gladly twist our necks to prolong the view whenever one speeds past. We should also not forget to mention the joy we feel when we're an "earwitness" to the grandiose sound of an Austin Healey at close range. The roar of its exhaust system is certain to elicit paroxysms of delight from every automobile enthusiast.

Nowadays, a well-organized and globe-trotting coterie of Austin-Healey owners keeps this myth alive. Members hold meetings, organize excursions and races, participate in one or another concours d'élégance, and display their Austin-Healey motorcars at every appropriate occasion. Incidentally, this brand takes its name from a Briton -- Donald Mitchell Healey, who in 1945 founded the Donald Healey Motor Company, which first called admiring attention to itself as an automobile manufacturer with the debut of the Healey 100. Leonard Lord, the president of the British Motor Company, likewise noticed the new roadster and, on the evening before the public presentation of the Healey 100 in 1952, Lord and Healey agreed to closely collaborate under the aegis of the British Motor Company. The first twenty Austin Healey motorcars were built in Warwick, before production was relocated to Abingdon in 1957. Serial manufacturing of the brand's cars in the USA was suspended in 1967, in the wake of stricter safety and exhaust regulations, but production continued in smaller numbers until 1972.

VINTAGE RALLY
Front of Austin Healey 3000 MKII at the Healey event in USA (2011)

Practice exercises at one of the Frédérique Constant Healey Challenge events

Austin Healey first achieved widespread fame thanks to the rally driver Pat Moss, the sister of motorsport legend Stirling Moss. She drove a factory-made vehicle which she nicknamed "The Pig" because it was always unpredictable and temperamental, and because its rear end had a latent tendency to swerve. This character flaw was also one of the reasons why so many Austin Healey cars, which were never manufactured in very large numbers to begin with, landed in ditches alongside roadways. Then as now, the driver of an Austin Healey always had to steer his classic vehicle with liberal sensitivity and without an unduly heavy foot on the accelerator pedal.

Peter, why did you form partnerships with several Austin Healey clubs?

PETER: Frédérique Constant knows how extremely passionate dyed-in-the-wool Healey enthusiasts can be about their cars. With our "Live Your Passion" slogan in mind, it simply made sense for us to combine our own passion for mechanical timepieces with another passion: namely, the love for special automobiles, an amour that's regularly celebrated at various international meetings. The cooperation between Frédérique Constant and the Healey brand is advantageous for everyone involved, so we and Healey are also planning other interesting activities for the future.

Frédérique Constant sponsored a very special event in 2008 the "Third European Healey Meeting" in Halmstad, Sweden.

Enjoying the ride at the Frédérique Constant Healey Challenge USA (2009)

PETER: The owners of more than 250 Austin-Healey sports cars from the early 1950s came to Sweden for the meeting. Approximately 30 particularly unusual models could be admired at first hand. All in all, more than 500 Healey enthusiasts from 18 different countries participated in the meeting. Healey fans from Australia journeyed the furthest to attend the event. In and around Halmstad, participants took advantage of opportunities to drive their automobiles, to compete with one another and to intensively celebrate their mutual passion for Austin-Healey. Never before had an international event brought together so many Healey models, including many of the most unusual.

As chief sponsor of many Healey events, Frédérique Constant has repeatedly debuted numerous special watch models.

PETER: We create a limited "Healey" edition whenever Frédérique Constant serves as chief sponsor of a Healey meeting. Aficionados can choose among several different variants and we limit each series to preserve its exclusivity. Of course, as a genuine goodie and eye-catcher, we deliver each watch together with a high-quality miniature model of an Austin-Healey motorcar in a wooden box.

The Frédérique Constant watch brand, which is strongly affiliated with automotive sport, found another interesting field of activity at international rallies. The Carrera Panamericana is one of the last remaining high-speed road races. Local police keep ordinary traffic and pedestrians safely off the roads while the vehicles traverse more than 3,000 kilometers across Mexico. Police cars actually accompany the roadsters

Austin-Healey

VINTAGE RALLY
Vintage Rally Healey Chrono
Automatic FC392CH6B4 (2009)
Limited edition of 1883 pieces

over large stretches of the route. The Carrera Panamericana began in the 1950s. After the Mexican portion of the Panamericana highway was completed in 1950, the Mexican government organized a five-day road race that spanned the entire country. The event was intended to attract international businesses to Mexico and to serve as an advertisement for the road, which was a masterpiece of engineering in its day. The racecars drove at a very fast pace along the route, which led almost entirely along the new highway that traverses the nation from north to south over a total distance of 3300 kilometers. The legendary race took place four times before it was ultimately called off. (1)

Thanks to the efforts of a group of Americans and Mexicans who were united by their mutual enthusiasm for motor sports, the Carrera Panamericana celebrated its revival in 1988. The route was divided into "transit" and "speed" sections. The participating vehicles drove together with ordinary road traffic during the "transit" part of the race, when all participants were obliged to obey traffic laws. The rally's drivers were then permitted to put the pedal to the metal in the "speed" section of the race, for which Mexican police cleared individual stretches of roadway. The participants started at thirty-second intervals and could really have fun. There's a theoretical speed limit of 144 mph or 232 kmh, but who would dare to drive so fast in a historical vehicle and on sometimes outrageously curvy roads? The fact remains that the Carrera Panamericana is one of the last remaining races where drivers are allowed to travel at such high speeds – and to do so under the watchful eye of the police!

(1)

Frédérique Constant was Official Timekeeper of the La Carrera Panamericana (2011)

La Carrera Panamericana.com.mx
TELMEX

ZACATECAS
AGUASCALIENTES
GUANAJUATO
QUERETARO
MORELIA
PUEBLA
OAXACA
HUATULCO
Oct.
2011

371

VINTAGE RALLY
Collage of photos from the La Carrera Panamericana (l), FC-435S6B6 (r)

OFFICIAL
TIMEKEEPER
FREDERIQUE CONSTANT
GENEVE
OFFICIAL
TIMEKEEPER
START
ITAL
U -5
GENEVE

(1)

Peking to Paris logo, Frédérique Constant was Official Timekeeper (2010)

Peter, why did you opt to also sponsor this daring road race?

PETER: A few years ago, we decided to expand our activities into the field of classic car rallies. Our first sponsorship was for the famous Peking to Paris rally (1). At the same time, we gave the name "Frédérique Constant Vintage Rally" to our new collection of classic car watches. Our partnership with the Carrera Panamericana fosters this still relatively young collection of watches. In the future, we'll be part of other classic-car rallies and thus continue to actively promote our Frédérique Constant Vintage Rally design. I'm convinced that classic cars and Frédérique Constant go very well together. Our customers appreciate classical design, which equally distinguishes vintage motorcars and our wristwatches. Furthermore, automobiles and watches both reflect a passion for mechanisms, which is shared by men around the world. .

The Peking to Paris rally is perhaps even a bit crazier and simultaneously also much more adventurous. The most recent repetition of this famous race occurred in 2010 and was aptly entitled "Peking to Paris – Driving the Impossible." Its 107 participants and their historical vehicles attempted to travel a route from Peking, along the Great Wall of China, through Inner Mongolia, the Gobi Desert and Outer Mongolia, to Ulan Bator, through Kazakhstan to Tashkent, onward through Uzbekistan to Samarkand, from there to Turkmenistan, into Iran to Tabriz, into Turkey to Istanbul, then through Greece, Italy and France to Paris. That's a total of 14360 kilometers or 8923 miles! The oldest automobile in the rally was built in 1922 and the most recent one was manufactured in 1954. Incidentally, the route followed the historical Silk Road so the drivers could enjoy their well-earned days off in fabled cities such as Almaty and Samarkand.

VINTAGE RALLY
Start of the Rally in Peking 2010

page 182
VINTAGE RALLY
Ankunft der Rallye "Peking to Paris" auf dem Place Vendôme in Paris 2010

FREDERIQUE CONSTANT
GENEVE
Peter Little

since 1907
PEKING TO PARIS

VINTAGE RALLY
Vintage Rally "Peking to Paris"
FC-393S6B6
Limited edition of 1888 pieces
(2011)

This monster of a rally traces its historical roots to 1907, when it was first organized. The idea for the race was born when the Parisian daily newspaper "Le Matin" published a daring challenge on January 31, 1907. It read, "Now the task is to prove that a man with an automobile can do whatever he wants and drive wherever he wants. Would anyone accept the challenge of driving an automobile from Peking to Paris?"

No fewer than ninety years passed before the second Peking to Paris rally took place. The next staging occurred in 2007 and the race will now be driven more often and more regularly. Incidentally, the next running of the rally is scheduled for 2013, when "Peking to Paris 2013 – Driving the Impossible" will, of course, begin in Peking. Other points along the way will include the Great Wall of China, Inner Mongolia, the Gobi Desert, Outer Mongolia, Ulan Bator, Khovd, Russia, Novosibirsk, Omsk, Tyumen, Samara, Ukraine, Kiev, Poland, Krakow, Bratislava, Salzburg, Gstaad, France and Paris.

That will be a genuinely superlative event, Peter, and Frédérique Constant will be its official timekeeper.

PETER: Under our motto "Live Your Passion," Frédérique Constant inspires people around the globe to do just that, i.e. to live their passion! Our partnership with the great Peking to Paris rally perfectly reflects this commitment. All the vehicles that will participate in the upcoming rally will be marked with Frédérique Constant's logo. When they arrive in Paris, the pilot and copilot on the winning team will each receive an exclusive timepiece from our Vintage Rally collection as a special gift.

VINTAGE RALLY
Frédérique Constant logo on all participating cars Peking to Paris (t), Giftbag Peking to Paris (rb), Healey participant Healey Challenge USA (rb)

FREDERIQUE CONSTANT
GENEVE
Live your passion

26
FREDERIQUE CONSTANT
GENEVE
Live your passion

044
FREDERIQUE CONSTANT
GENEVE
PPIHC

Alpina
BIENNE,
Quai du Haut 51
4 Janvier

ALPINA

FROM BUYING SYNDICATE TO INNOVATIVE WATCH MANUFACTURER

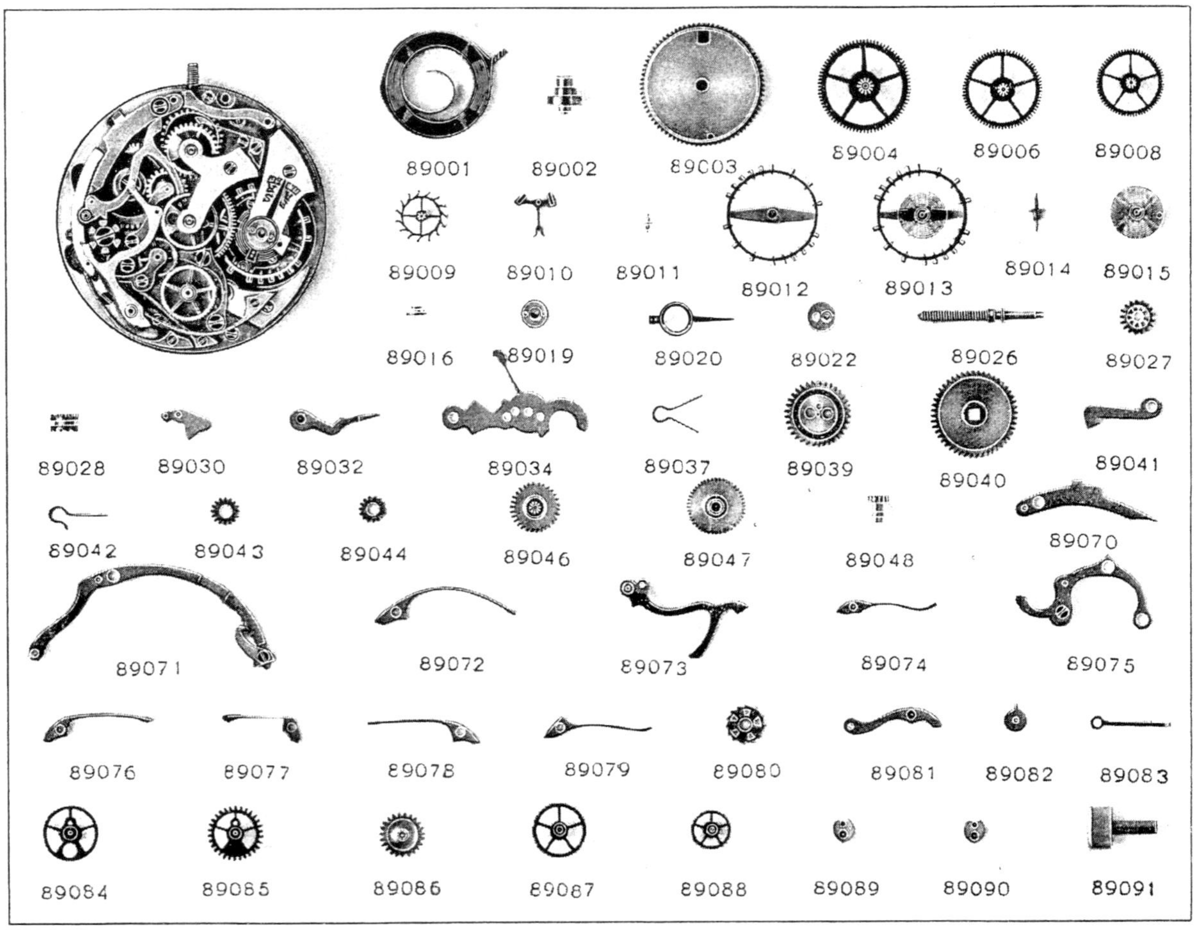
89001
89002
89003
89004
89006
89008
89009
89010
89011
89012
89013
89014
89015
89016
89019
89020
89022
89026
89027
89028
89030
89032
89034
89037
89039
89040
89041
89042
89043
89044
89046
89047
89048
89070
89071
89072
89073
89074
89075
89076
89077
89078
89079
89080
89081
89082
89083
89084
89085
89086
89087
89088
89089
89090
89091

(1)

GOTTLIEB HAUSER
Founder of the "Union Horlogère"

PROLOGUE

Victor Hugo's realization that "nothing is more powerful than an idea whose time has come" aptly applies to an idea that occurred to Gottlieb Hauser. Two years before the French author of the aphorism ended his life, Hauser, a watchmaker in Winterthur, founded in the context of the national exhibition in Zurich in 1883 an association which would later become the source of the Alpina watch label. The constitutional assembly took place in Zurich's Kronen Hall on May 8, 1883.

But we're getting ahead of our story. A deeper background was behind the founding of the "Schweizerische Uhrmacher-Corporation" (i.e. "Swiss Watchmaker Corporation") in the late 19th century. During this epoch, more and more consumer cooperatives and department stores were increasingly gaining footholds in the marketplace. Small and midsized retailers were particularly hurt by the stronger competition. Ongoing competitive battles called for countermeasures, e.g. communal purchasing of merchandise in accord with the idea that unity confers strength. By joining together to purchase goods, the syndicated businesses could exert greater pressure on suppliers and negotiate more favorable prices. The members stood behind these purchasing syndicates as principal equity investors, so the legal form of the cooperative association played an important role.

A FUNDAMENTAL RETROSPECTIVE

Our look at 1883 almost necessarily calls for elucidation of the watch industry in that era, when brands per se were still largely unknown. Legal regulations to protect them first took effect in Switzerland on January 1, 1880. The dials of watches were identified with the names of the manufacturers or the specialized dealers who purchased and sold the products. The contemporary brand name watch as such did not yet exist. On the other hand, the watchmaking industry was inspired by American achievements to gradually adopt mechanized manufacturing methods, which went hand in hand with larger numbers of units produced, greater precision and simpler interchangeability of parts for servicing and repairs: This reduced the need for costly modifications formerly required to coax new components into places vacated by old ones. But the industry wasn't quite so far advanced in 1883. Before a watch was sold, it typically had to be disassembled, the meshing of its gear-train and other parts of its movement had to be checked and, if necessary, everything would have to be reworked. Watchmakers and dealers also found it costly and cumbersome to keep spare parts in stock. Furthermore, it wasn't always easy to judge the quality of purchased items.

After repeatedly encountering such problems, Gottlieb Hauser (1) took action in 1883 by establishing the "Vereinigung der Schweizer Uhrmacher" ("Association of Swiss Watchmakers"), for which he defined two essential objectives:

1. To seek manufacturers who could supply reliable high quality watches and accessories to watchmakers, and

2. To negotiate favorable conditions as a consequence of appealing group orders. In this way, the participating watchmaker would be able to efficiently influence the market. This initiative among retailers was completely new.

There was also another irksome situation. Smaller manufacturers and dealers paid frequent and sometimes rather obtrusive visits to watchmakers, thus robbing their hosts of valuable time at the worktable. This needed to be stopped.

The participants at the inaugural meeting unanimously chose Gottlieb Hauser as the association's president. Hauser's plan was for the watchmakers to initially concentrate on collective purchasing. Potential suppliers didn't yet trust the new organization, so its president was had to personally vouch for 4000 Swiss francs in 1883 and once again in 1884. Hauser also initiated measures to form national trading companies in Switzerland and abroad. His foresight prompted the founding of similar organizations in Germany, Austria and the Nordic countries. In the meantime, the "Alpina Principle" had undergone considerable evolution. Progress in the manufacturing sector went along with commercial aspects. Hard work and consistently cooperative policies paid off. The 1884-1885 fiscal year ended with a balance of 443 francs.

SUCCESS STORY: PART ONE

The idea, whose time had indeed come in 1883, gathered momentum. For legal reasons, the association was renamed the "Schweizerische Uhrmacher-Genossenschaft" ("Swiss Watchmaker's Cooperative Association") or "SUG" in 1886. Its bylaws stated: "Any watchmaker who resides in Switzerland can become a member. Applications must be submitted in writing and accompanied by a recommendation. Applicants' names will be made known to all members. If no objections are raised within a fortnight, the chairman will accept the applicants into the syndicate. Along with an entry fee, which shall initially be set at five francs, each member must pay 500 francs as a security deposit.... Members are excluded from personal liability. Only the syndicate's assets shall be liable for the association's liabilities. Every member has the right to avail himself of a merchandise credit with a maximum value equal to that of his deposit."

THE CONCEPT AND BYLAWS MUST HAVE BEEN CONVINCING, BECAUSE INCREASINGLY MANY WATCHMAKERS JOINED

In collaboration with qualified manufacturers, the development of the association's own movements and the establishment of a distribution network began in 1883. Movement-blanks were initially produced by Duret & Colonnaz, a Geneva-based ébauches factory. From the start, the association established a collaborative arrangement with a manufacturer of complete watches in Biel. Blanks were given their finishing touches and assembled into complete watches at Straub & Cie., which was directed by Jakob Straub. Too few qualified workers were available in Biel, so "good workers from French-speaking Switzerland" were also hired.

The "Vereinigung der Schweizer Uhrmacher" was thus able to offer its members meticulously manufactured timepieces of progressively better quality at genuinely attractive prices. Thanks to the competence of watchmakers from western Switzerland, difficult repairs could also be performed in the association's ateliers.

(1)

Alpinist meeting

NOT EVERYTHING RAN SMOOTHLY

Agreements with the two manufacturers specified that the syndicate's orders would be large enough to keep production at full capacity. To achieve this, all members were obliged to generate at least half of their revenue via orders placed through the syndicate. Despite its many positive side-effects, the concept had an Achilles' heel. Not all members abided by the agreements, so the holding company was unable to order enough merchandise. Manufacturers complained that the syndicate had failed to comply with its daily obligations to order two cartons, each containing 144 watches. This dilemma arose because the syndicate's purpose was to manufacture and sell only good quality pocket-watches, and its members were expected to act accordingly. Calls to also include "cheap merchandise" were repeatedly heard, but the majority of members were unwilling to go along with this. The organization therefore agreed to also accept new members whereby the statutes granted guaranteed exclusivity to earlier members.

GROWTH

The contract with Straub & Cie. made it advantageous to relocate the syndicate's headquarters to Biel in 1890. The administration and the stockpile were centered here from that time forward. The first office consisted of four rooms, a small storeroom and a workshop.(1)

The headquarters also catered to more than 200 German customers via a subsidiary in Konstanz. Bookkeepers reported total revenues of 200,000 Swiss francs at year's end.

But the syndicate's success was a thorn in the eye of other manufacturers and wholesalers, who repeatedly sought to harm the association and its house periodical. Especially through the association's newsletter, new members could be attracted who had not previously heard anything about the existence of a Swiss watchmakers' association.

The Swiss watchmaking industry suffered a crisis and earned significantly less revenue than usual in 1891. Ebauche manufacturers undercut one another by selling their wares at dumping prices. The fast-growing German market was a welcome sunray in this otherwise bleak situation. Becker & Cie. in Frankfurt acquired the German subsidiary in 1892. It evolved into the general representation, with its headquarters in Germany's capital city of Berlin, in 1899. Business continued on an upward trajectory.

THE WAY INTO THE 20TH CENTURY

The SUG was permitted to participate in the Parisian World Exposition in 1900 under the condition that it sold no merchandise to private individuals. Dealers were expected to uphold this policy, but most other exhibitors defrayed their expenses by selling to the public.

Straub & Cie. unveiled a special caliber around the turn of the century. Commercial progress and an increase in revenue to 300,000 Swiss francs necessitated a move to more spacious quarters. A new building at Seevorstadt in Biel provided sufficient space. Straub's suggestion of establishing a "Société des fabricants d'horlogerie suisse réunis," i.e. an additional manufacturers' association, was eagerly accepted. A change in its statutes also gave a new name to the association, which would henceforth be known as "Union Horlogère, Schweizerische Uhrmachergenossenschaft, Association horlogère Suisse."

Gottlieb Hauser and his son went on an extended business trip to Russia in 1903. For fifty successive days, they strove to establish a foothold in the czarist realm. Their efforts succeeded: by end of their trip, they had entered into their ledgers the names of surprisingly many new customers – including some famous ones. The new clientele helped to increase revenues to circa half a million Swiss francs.
The question of establishing a general partnership arose in 1904. The business was founded and named "Union Horlogère, Uhrenfabrikation und Handelsgesellschaft, Biel, Glashütte, Genf." The 1905-1906 fiscal year was the first in which the company earned more than one million Swiss francs. A 49% increase over the previous year's earnings proved that there were ample opportunities for further growth for the company, which was also represented in Austria-Hungary, Belgium, Holland, Italy and Spain. Even overseas buyers had become members.

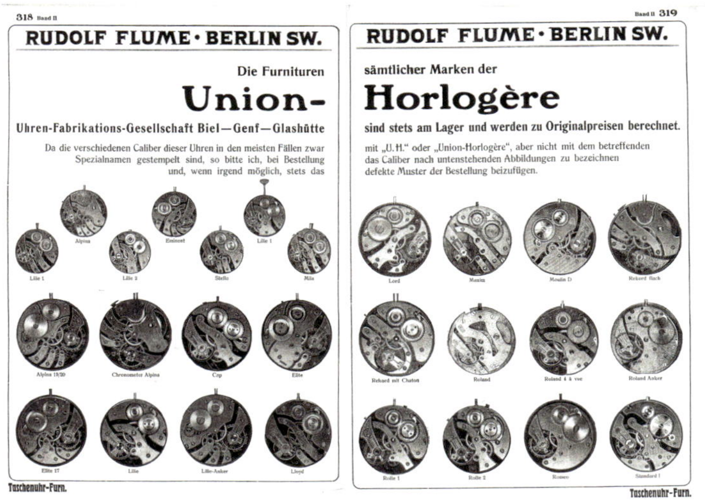

SUCCESS STORY: PART TWO

Success was likewise enjoyed by the Société des fabricants d'horlogerie suisse réunis. Some two dozen manufacturers benefitted from the association's orders. As the years passed, this circle came to include (among others) J. Straub & Co. in Biel (complete watches), Favre in Geneva (complete watches), Kurth Frères in Grenchen (complete watches – later Certina), Duret & Colonnaz in Geneva (ébauches), Huguenin-Robert (cases), Schwob Frères & Co in La Chaux-de-Fonds (Cyma – complete watches), Robert Frères in Villeret (complete watches – Minerva), the Ali Jeanrenaud factory for fobs and bows, and the Numa Nicolet et Fils dial factory.

Another step forward came with the establishment of a comprehensive international guarantee originally offered by the society in 1906. Its members vouched for the proper functioning of the watches, thus underscoring their solidarity and the trustworthiness of their products. The mutual guarantee, however, covered only timepieces that had been produced by the Union Horlogère.

NEW MEASURES FOR THE FIRM'S 25TH ANNIVERSARY

When the prospering business celebrated its 25th anniversary in 1908, the festivities also included the registration and subsequent protection of the name "Alpina" as a watch brand for high-quality calibers, which had been produced since 1901 and frequently included Breguet hairsprings and compensating balances. This step was urgently required, because after the turn of the century, more and more manufacturers had adopted short and catchy names that promised greater success in advertising and sales. To uphold the prestige associated with the new name and after a lengthy discussion, the members agreed to use the name "Alpina" only for a dozen selected calibers and for timepieces that encased these movements.

As was de rigueur in this epoch, a catchy name necessarily also needed a distinctive logo, which was duly created in the form of a red triangle with a stylized dial and the brand's name. (1)

Later on this eye-catching signet unmistakably identified exclusive Alpina dealers. One of the association's principles was, that wherever possible, there would be only one member in each town. Only in larger cities was this guideline occasionally waived. The communal advertising measures, which no individual could finance on his own, also included relatively lavish catalogues. Suppliers were expected to contribute toward defraying the expenses of these publications, although there were a few exceptions to the rule. The program also included collectively purchased newspaper advertisements.

But merely having many members under one roof didn't necessarily mean that they would all think alike and speak with one voice. Different people had different opinions about the advantages associated with an Alpina concession. This was also true for Alpina, the Swiss watch company, as one of the common denominators of the Alpina watchmakers' organizations. The emphasis on cooperative and coordinated activities was widely accepted in Germany, Austria, Holland, Denmark, Norway, Sweden and Finland, but watchmakers in other countries were less keen about what they regarded as all too close collaboration and its attendant obligations. Here, Alpina watches found their way into dealerships through the traditional route via general representatives. The ALPINA Union Horlogère watch factory, which had been newly opened by Straub & Cie., earned annual revenues totaling two million Swiss francs. (2)

(1)

ALPINA LOGO
The triangle form
Iconic signature of Alpina

(2)

ALPINA
UNION HORLOGÈRE

Alpina
Alpina

MANUFACTURE
Union Horlogère Manufacture in Bienne

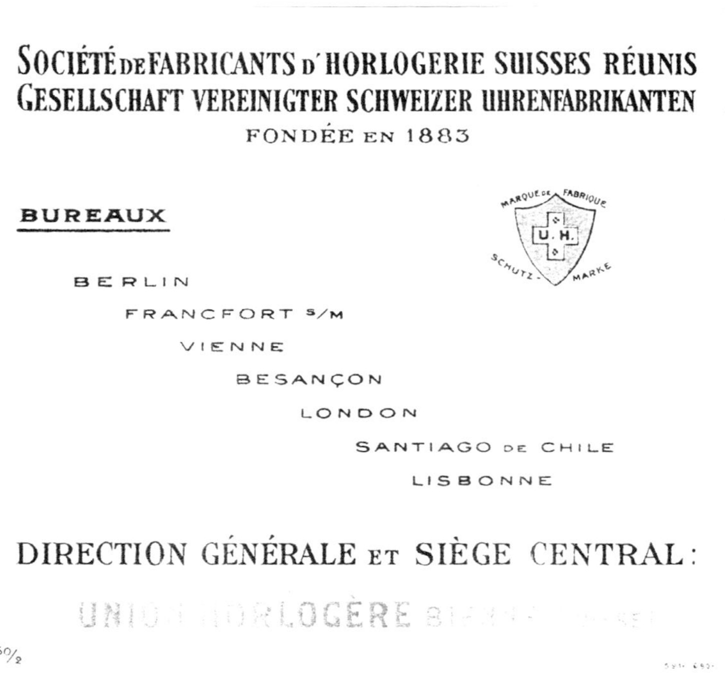

SOCIÉTÉ DE FABRICANTS D'HORLOGERIE SUISSES RÉUNIS
GESELLSCHAFT VEREINIGTER SCHWEIZER UHRENFABRIKANTEN
FONDÉE EN 1883

MARQUE DE FABRIQUE U.H. SCHUTZ-MARKE

BUREAUX

BERLIN
FRANCFORT S/M
VIENNE
BESANÇON
LONDON
SANTIAGO DE CHILE
LISBONNE

DIRECTION GÉNÉRALE ET SIÈGE CENTRAL:

UNION HORLOGÈRE

50/2

HISTORIC TIMEPIECES
Alpina pocket watches

HOW THE UNION HORLOGÈRE FUNCTIONED

Pursuant to its bylaws, the Union Horlogère focused its efforts on high-quality watches. Beginning in 1908, most of these timepieces bore the Alpina insignia. Every watchmaker, manufacturer or dealer who wanted to join the Union was obliged to request membership, whereupon the chairman would subject the candidate to careful scrutiny. Acceptance or rejection of the applicant was decided in consultation with other members. After a new concessionaire had paid his membership fee, he could purchase Alpina watches at favorable prices.

Other advantages included comprehensive efforts by the organization on behalf of its members, who were free to sell products made by other brands but not allowed to join similar organizations. Dealers were also expected to sell the merchandise at the prices that had been set for their country and were prohibited from undercutting fellow dealers by offering discounts. The expenses of all advertising for Alpina watches were paid from the Union's treasury, which earned its income primarily from subscriptions, from fees paid by new members, and from rewards depending on total revenues. The guarantee association that had first been introduced in 1908 was successively broadened to include all Alpina dealers. It became valid throughout the world in 1926. Concessionaires could also rely on the association's window-dresser to create attractive displays in their shop windows. Regularly scheduled sales and technique training sessions helped to increase revenue, as did the association's newspaper, which kept members informed about the latest developments and news.

The association was directed by a committee that convened several times each year. The topics on its agenda included presentation of new patents, clarification and settlement of disagreements among members, selection of the catalogue's contents, and the determination and distribution of costs. An important event was the annual congress, which was also a small watch fair where the so-called "Alpinists" could view and order new products, discuss relevant issues with their peers, and cultivate amicable contacts with one another.

THE CHAOS OF WAR

Needless to say, the Union Horlogère was not unaffected by the outbreak of the First World War on August 1, 1914. Switzerland's declaration of neutrality and its standing army kept the Alpine federation out of the conflict, but commerce nonetheless suffered from the war. A heavy blow struck the Union Horlogère when the Entente put it on a blacklist after it had invested capital in the Triple Alliance. A ray of sunshine was the conferral of a gold medal at the Swiss National Exhibition in 1915. The following year, both parts of the business – i.e. administration and production – remained under the close scrutiny of the Entente nations. The war and its adverse consequences accelerated an unavoidable regrouping so that the Union would be able to regain a commercial foothold. This included reorganizing the headquarters, which until then had been organized as an association, into a joint-stock company. The German branch became an independent association. This decision was made at an extraordinary general meeting on February 12, 1917.

Shares of the Swiss central division were converted into bonds of the new joint-stock company. Otherwise the Swiss enterprise did business under the name "Union Horlogere, Schweizerische Uhrmacher-Genossenschaft," while the "Alpina Deutsche Uhrmacher-Genossenschaft G.m.b.H. Berlin" continued to do business as an independent company which developed into a surprisingly strong organization after the war. The urgently necessary liquidation contract was signed on November 25, 1916. Among its other stipulations, right to also use the established brand was granted by the Swiss Union Horlogère A.G. to the newly founded German company. Misuses were categorically prohibited and a licensing fee of two percent was to be paid to the Swiss. A unifying element was the establishment of a common central office in Biel. Despite its restructuring, the company remained on the Allies' blacklist in 1917 and 1918. Revenues in Switzerland were low and assets were frozen in England, where a solvent company purchased the inventory.

Alpina Chronometre produced in Glashütte

Another joint-stock company was nonetheless established on September 12, 1916. The new company was headquartered in Biel, placed on the Allies' blacklist, named Meteore A.G. and founded with the purpose of producing, buying and selling products for the watch industry. A temporary way to avoid prohibitions associated with radium trading was found by establishing the "Radio-Chimie Bern" public company, in which neither the Union Horlogère A.G. nor the Meteore A.G. participated financially. The company was renamed "Radiochimie Meteore A.G." in 1921. As the business' name clearly stated, it continued to work with radioactive material to make luminous dials, for which demand was steadily increasing.

An unprecedented continuation of the success story began at the end of the First World War. About 2,000 dealers throughout Europe sold Alpina watches, which could be purchased in Lisbon, Amsterdam, Copenhagen, London, Stockholm and even Santiago de Chile.

Unlike its successes in Europe, however the attempt to gain a foothold in North America was a dismal failure. The excursion across the Big Pond nearly ended in financial disaster. Fortunately, the importer of a large shipment of watches didn't pay duty fees at the customs office, so the delivery was returned to Switzerland, where the manufacturers suffered bruised pride but no lasting damage.

ALPINA AT GLASHÜTTE IN SAXONY

Watchmaking in Glashütte has traditionally been characterized by thriving coexistence and symbiosis among major manufactures and very small, individually operated ateliers, but the situation was entirely different for the "Präcisions-Uhrenfabrik Alpina," which was founded at Glashütte in Saxony in 1909. Its genesis was derived from the effort to advance into the elite segment of absolute chronometric luxury. Partly because of its brief history, Alpina was unsuitable for this venture, so the decision-makers opted for Glashütte, the recognized Mecca of the German art of watchmaking. The goal of the company, which had charter capital valued at 20000 Goldmark, was "to manufacture precision watches of the best quality and primarily relying on the caliber which is protected under the name 'Alpina.'" The most important watchmaking company in Saxony's Müglitz Valley at this time was A. Lange & Söhne. Anyone who wanted to compete with it would have to produce merchandise in a quality that was at least on a par with Lange's. A 19-ligne caliber (diameter = 42.86 mm) was obviously well suited for this. The "Chronometer Alpina" with Swiss lever escapement had thoroughly proven itself for many years. It was revised for the Saxon version to include a Glashütte lever escapement. High-quality gold coating was applied to the movement's surfaces on the premises. Fine adjustment was performed in Switzerland by W. A. Dubois, a recognized master of his métier. Bearing the words "Präcisions-Uhrenfabrik Alpina Glashütte i. S." on their dials and movements, these comparatively costly products first appeared in dealers' showcases in 1912 and were sold together with guarantee certificates and lavish etuis. A letter circulated in November 1912 included the following prose: "Kindly permit us to explain the advantages of our Alpina Glashütte, which are manufactured entirely in Glashütte and encase our Alpina ébauches from Geneva. The gold escapement is an original Glashütte product of the very finest quality. The wheels, jewels, etc. for our watches are likewise made exclusively in Glashütte. Our Alpina Glashütte watches therefore deserve to be regarded not only as genuine, but also as first-class Glashütte products. They are completed in only one best version and are qualitatively equal to the finest merchandise offered by other manufacturers in Glashütte.... We should also not fail to mention that our factory in Glashütte can sell its products at significantly lower prices than those charged by A. Lange & Söhne and can also offer you correspondingly greater usefulness. It is understandable that due to clients' demand, some of our members also keep the well-known watches from A. Lange & Söhne in their stock, but other Glashütte manufacturers, as well as second-quality merchandise from A. Lange & Söhne, will not be accepted into our members' stock." A 21-ligne caliber served as the basis for an additional model in 1913. Intended for use by the navy, its dial, hands and case were installed in a special atelier in Glashütte with assistance, depending on the amount of work pending, by watchmakers who worked in their own homes.

(1)

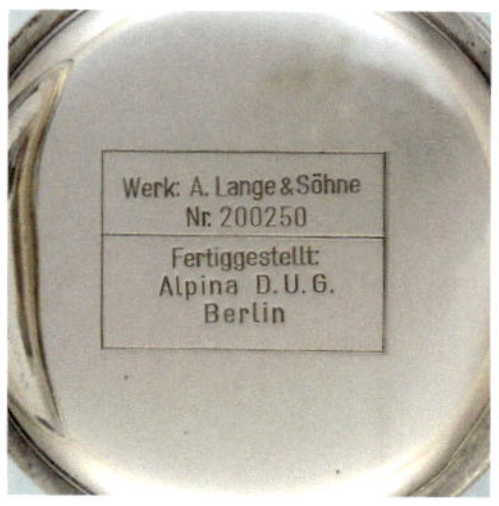

Alpina with Lange & Söhne caliber

JUDICIAL RULING

A. Lange & Söhne gave its competitor a decidely icy reception. The traditional German manufacture sued Alpina and the Assmann company in 1913 at Lauenstein District Court, where Lange insisted that its competitors had willfully misused "Glashütte" as a designation of provenance. But a significant percentage of the value of the products resulted from work performed in the Müglitz Valley, so the plaintiff's arguments failed to convince local entrepreneurs, the community and its superior authority. Georg Heinrich, who had directed the Assmann firm since the death of its founder Paul Assmann in 1911 and who had been employed by Lange for many years, mustered persuasive arguments in favor of Alpina: "During the tenure of my employment at Lange, it seemed quite reasonable to me that the firm would order types of movements which, for financial or other reasons, could not be manufactured on Lange's own premises and would purchase them from Swiss suppliers who were extraordinarily highly developed specialists in the fabrication of ébauches. It also seemed eminently logical that other Glashütte-based companies would follow Lange's example, initially for purchases of complicated mechanisms." (1)

Caught in the middle of this legal battle was Professor Ludwig Strasser, the director of the German School of Watchmaking, who had been tasked with drafting an objective appraisal, on the basis of which the court ruled against the plaintiff in 1915. But the "Präcisions-Uhrenfabrik Alpina in Glashütte" gained little from this courtroom victory. A war was raging and people had other, more pressing concerns than to purchase luxurious watches, no matter where those timepieces may have been manufactured. As a whole, this chapter in Alpina's history was not crowned with success. Probably only a few hundred "Glashütte Chronometer Alpina" were sold to consumers. Press reviews that were not always positive further discouraged sales, and wartime trade restrictions likewise put a damper on selling. The company was disbanded on July 17, 1922.

U.H.

PRÄZISIONS-ANKERUHR
ALPINA-GLASHÜTTE

No [redacted]

UNION HORLOGÈRE

Biel = Glashütte i/Sa. = Genf

Diese Uhr ist aus bestem Material unter Verwendung des bewährten Glashütter Original-Gold-Echappements allererster Qualität – compensierender Chronometer-Unruhe, harter isochronischer Spiralfeder, Zapfenlöcher aus feinstem Rubin – angefertigt.
Jede Uhr ist in allen Lagen und Temperaturen auf das Genaueste reguliert und wird für absolute Zuverlässigkeit des Ganges unbedingt garantiert.

UNION HORLOGÈRE.

Alpina Union Horlogère Guarantee Certificate

THE PATH INTO A NEW ERA

Article 2 of the cooperative association's bylaws was revised in 1928. The new version stated: "The association's purpose is to fabricate watches and related articles, to trade with them, and to encourage solidarity among industrious watch specialists and retailers in order to offer them nonmaterial and material advantages." After many successful years of service to the syndicate, Gottlieb Hauser and Jakob Straub submitted their resignations from the board of directors. Their successors were Gottfried Peter (who directed a bank in Biel) and Louis Straub (who manufactured watches there).

The first task for the new board members was to hammer out the details of a cooperative venture, first proposed in 1927, between the association and the Gruen Gilde watch factory. Under no circumstances would the collaboration be allowed to tarnish the Alpina brand name, so a long and profound discussion began. These talks initially failed to achieve concrete results, but the board of directors finally voted in favor of the venture on September 8, 1928. An extraordinary general meeting had to be called. The assembly, which convened on October 8, 1928, first debated the necessary provisions and other measures, and then voted to implement them. The president of the board of directors submitted a detailed exposé because plans called for revising the structure of the enterprise and increasing its share capital to 1.5 million Swiss francs, but acceptance of the proposals become unnecessary when the contract was signed with Gruen.

ALPINA-GRUEN GILDE UHRENFABRIKEN BIEL (SCHWEIZ)

Alpina-Gruen Guilde
Fabriques d'horlogerie, Bienne (Suisse)

Wir haben von diesem Gruppenbild unserer Fabriken Separatabzüge machen lassen, und wer weitere Exemplare zum Aufziehen und Einrahmen wünscht, wolle es uns mitteilen.

Alpina-Gruen Guilde
Fabbriche di orologi, Bienne (Svizzera)

Nous avons fait faire de cet ensemble de fabriques des reproductions spéciales. Les intéressés qui désireraient recevoir un ou deux exemplaires pour les encadrer sont priés de nous le faire savoir.

A SHORT BIOGRAPHY OF GRUEN

Dietrich Gruen initially made only ten watches each day in his little atelier. The year was 1874 when Gruen, a watchmaker of German extraction, began plying his trade in Columbus, Ohio. Prior to going into business as an independent artisan, he had completed an apprenticeship in Germany and afterwards made watches in Switzerland for three years. From 1879, his business was named "Gruen & Savage." Gruen's son Friedrich joined the little company in 1894 and the name was changed to "D. Gruen & Sohn." During this time, Gruen's watches encased calibers that had been made in Switzerland and Glashütte. The business's name was changed to "D. Gruen & Söhne" three years later in recognition of the fact that Gruen's second son, Georg J. Gruen, had joined the business as a partner. After the turn of the century, all of the movements that Gruen encased were of Swiss provenance. This fact was further reinforced with the establishment of the subsidiary "Gruen Watch Manufacturing Company" in Biel in 1903. As before, cases were made and watches were assembled in the USA, which was the largest and most important market for Gruen's merchandise. This arrangement exempted him from paying the high import duties which were levied on imported Swiss watches. In the meantime, Gruen had also begun designing and building his own hand-wound calibers. These included the "very thin" types, so named because of their exceptional slimness. Gruen launched his first wristwatches for ladies in 1908 or thereabouts. This new type of watch was massively advertised in the USA starting in 1910. To celebrate its 50th anniversary, Gruen's firm presented a wristwatch with a solid gold movement.

(1)

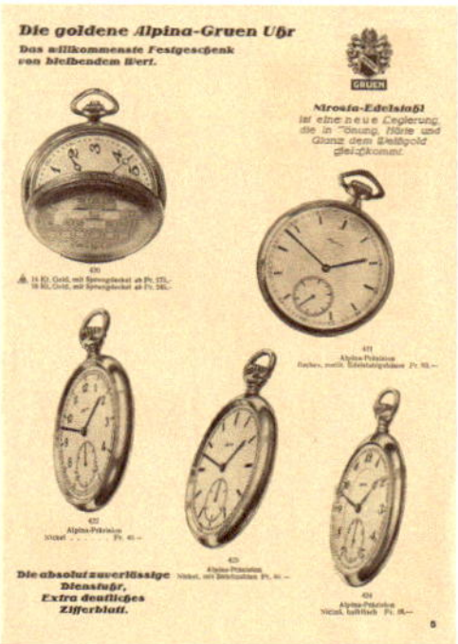

Damalige Alpina-Werbung
Earlier Alpina publicity

GRUEN, AEGLER AND ALPINA

One of the most important suppliers of movements in this epoch was Aegler, which did business under the name "Manufacture des Montres Rolex & Gruen Guild A." This also explains why identical calibers ticked inside the Rolex "Prince" and the Gruen "Duo-Dial" models in the early 1930s. Designated as caliber 877, the baguette-shaped, 16.6 by 32.7 mm, hand-wound movement provided 58 hours of uninterrupted running for the Alpina-Gruen-Tecno. In the Rolex "Prince," on the other hand, the ticking life was named "T.S." The Alpina-Gruen watch factories advertised this top-of-the-line model with a special brochure and the following words: "The new chronometer wristwatch boasts a trailblazing movement and shows the exact time in seconds: for physicians, engineers, technicians, transit officials, merchants and athletes. If desired, each Tecno will be delivered with an official rate bulletin." This special item necessary demanded a financial tribute from would-be owners. An ordinary rectangular stainless steel wristwatch encasing a shaped movement called for an investment of 95 francs in 1932, but buyers who were fanatic about precision had to part with at least 135 francs for the steel Tecno, a product of the "Alpina Gruen Gilde SA," which was officially registered in 1929. (1)

The marriage of Alpina and Gruen on May 13, 1929 was fated to be relatively short-lived. Alongside Alpina's chief, the board of directors also included Fred G. and George J. Gruen (industrialists in Cincinnati, Ohio), Albert Weber (an industrialist in Geneva) and Hermann Aegler (an industrialist from Biel).

The partners chose a lengthy but informative name for their enterprise, which was officially known as the "Alpina-Gruen Gilde Uhrenaktiengesellschaft (Union Horlogère) (Montres Alpina-Gruen Guilde Société Anonyme) (Union Horlogère) Alpina-Gruen Guild Watch Co. Limited) (Union Horlogère)." Headquartered in Biel, the company's purpose was "the manufacture of watches and related articles, as well as the wholesale selling of same, and the syndicating of watch manufacturers and retailers in order to offer them non-material and material advantages."

RECTANGULAR WATCHES
Rolex Prince and Alpina-Gruen Tecno (early 1930s)

Die patentierte Krone La couronne brevetée

Alpina Gruen*

A Die patentierte Alpina Gruen* Krone gewährt **vollkommenen** Schutz gegen **Staub,** indem der geringste Zwischenraum vermieden wird.

A La couronne brevetée Alpina Gruen* offre une protection **absolue** contre la poussière, le moindre interstice étant supprimé.

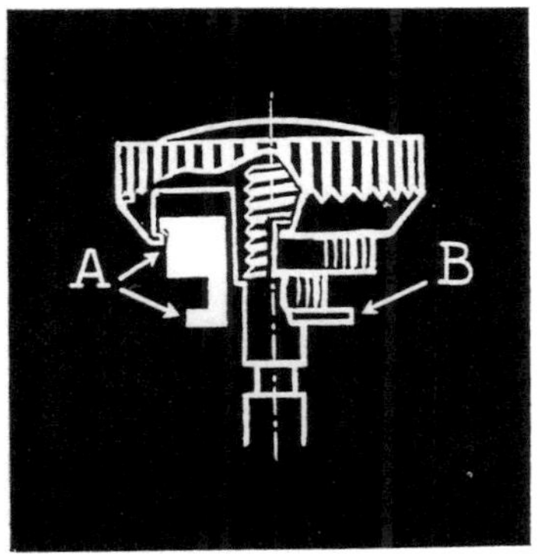

B Die patentierte Krone Alpina Gruen* kann weder verloren noch herausgerissen werden.

B La couronne brevetée Alpina Gruen* ne peut être ni perdue ni arrachée.

Verwenden Sie bei jeder Reparatur ausschliesslich die patentierten Alpina Gruen* Kronen. Bei Verwendung irgendeiner andern Krone ergibt sich ein freier Raum rings um die Aufzugswelle, durch welchen der Staub eindringen kann.

Pour toutes réparations, n'employez que des couronnes brevetées Alpina Gruen*. Tout autre genre de couronne, laissant un espace libre autour de la tige, favorise l'entrée de la poussière.

36 patentierte Alpina Gruen* Kronen, in unseren verschiedenen Kalibern sortiert, aus rostfreiem Edelstahl oder Goldplaqué, für Herren- und Damen-Armbanduhren; das Sortiment im abgebildeten Etui

Fr. 15.—

36 couronnes brevetées Alpina Gruen*, assorties dans les différents calibres, acier inoxydable ou plaqué or, pour montres bracelets hommes et dames; l'assortiment, dans l'étui illustré

Fr. 15.—

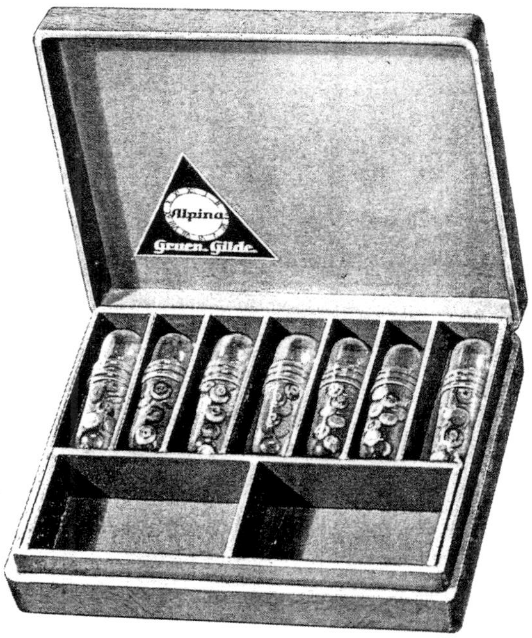

*) Alpina Gruen Krone, internationales Patent Nr. 1464.

*) Couronne Alpina Gruen: brevet international No. 1464.

PATENT
Alpina-Gruen patented crown

ADVANTAGES? WHAT ADVANTAGES?

But these "advantages" weren't very noticeable. Almost no one in Europe was familiar with Gruen, despite his high-quality watches and his partnership with Aegler, which was a partner of Rolex in Biel. Furthermore, Gruen charged significantly higher prices for his merchandise than the Alpina watch factory asked for comparable items, a fact which naturally didn't endear Gruen to European consumers. Where prices were concerned Fred Gruen claimed that Alpina-Gruen was the only corporate group which insisted on predetermined prices for its guaranteed high-quality products and which also pursued this pricing policy in America. Alpinists outside Switzerland weren't eager to follow suit. They complained that customers could often buy the watches for lower prices in Switzerland than in Alpinists' home countries.

Acceptance of the Gruen brand also suffered because the Americans obstinately refused to comply with repeated requests for financial contributions to the annual catalogue. Furthermore, Gruen hindered its partner Alpina from marketing its products in the USA. The marriage clearly wasn't a win-win situation. Recurrent quarrels precipitated painful losses. Conflicts with Rolex arose from time to time too. The bone of contention were concessions at locations where Alpinists were already active. In this context, Hermann Aegler said that wherever an Englishman asks to buy a Rolex, "this place thus becomes an alienated venue, but Rolex will, of course, not establish stores in small towns because it is Rolex's intention to serve only specialized shops." As far as it affected Alpina-Gruen Gilde, the modus vivendi was to permit sales "abroad." But in so doing, Alpina-Gruen Gilde had divorced itself from any obligation to Rolex. Lasting peace failed to ensue. Collaboration with Gruen and thus also with Aegler was terminated in 1935. The joint-stock public company that emerged from the divorce was named "Alpina Union Horlogère S.A." Its headquarters remained in Biel.

BRAND AFFAIRS

The "Schweizerische Uhrmacher-Corporation" had undergone several name changes, but fifty years after its founding, it had finally achieved strong acceptance with regard to distribution and product quality. The Alpinists were keen to preserve this good reputation, so the production of watches bearing this insignia remained the exclusive privilege of factories that belonged to the corporate group. In this sense, Straub & Cie. in Biel supplied all gents' pocket-watches and wristwatches, which encased either round or shaped movements. Favre in Geneva concentrated on timepieces for the fairer sex, whose watches likewise encased either circular or noncircular movements. Alpina-Gruen calibers were markedly costlier and of better quality than Alpina's. From 1929 to 1935, the Gruen Gilde brand represented the absolute top segment of the product palette. The temporary manufacturing of inexpensive watches bearing the "triangle-A" logo but without the Alpina insignia was totally discontinued. (1) The majority of the association's members refused to compromise, although calls for "accessible entry-level wares" were occasionally heard.

CALIBERS

In 1912, the spectrum of watches available from all of the brands which belonged to the Union Horlogère, Uhren-Fabrikations-Gesellschaft Biel-Geneva-Glashütte already included 27 calibers, all of which were circular in shape. Seven had diameters between 11 and 13 lignes, i.e. 25 to 29 millimeters. Most of these calibers graced ladies' pocket-watches, although some were also encased in wristwatches. The highest quality item was the beautifully designed "Alpina" caliber. Movements named "Alpina" were likewise the best ones among the twenty calibers for gentlemen's watches. The 19-ligne Alpina 19/20 was a so-called "Lépine" caliber with a subdial for the seconds at the "6." The equally sized Alpina Chronometer, which relied on savonnette construction and had its subdial

Alpina Pilot watch from 1930's

(1)

The Alpina Triangle Logo

HISTORIC TIMEPIECES
Alpina Pilot watch from the 40's

page 210-211
ALPINA MANUFACTURE
Technical drawing made by hand of Caliber 582/584

Alpina
943

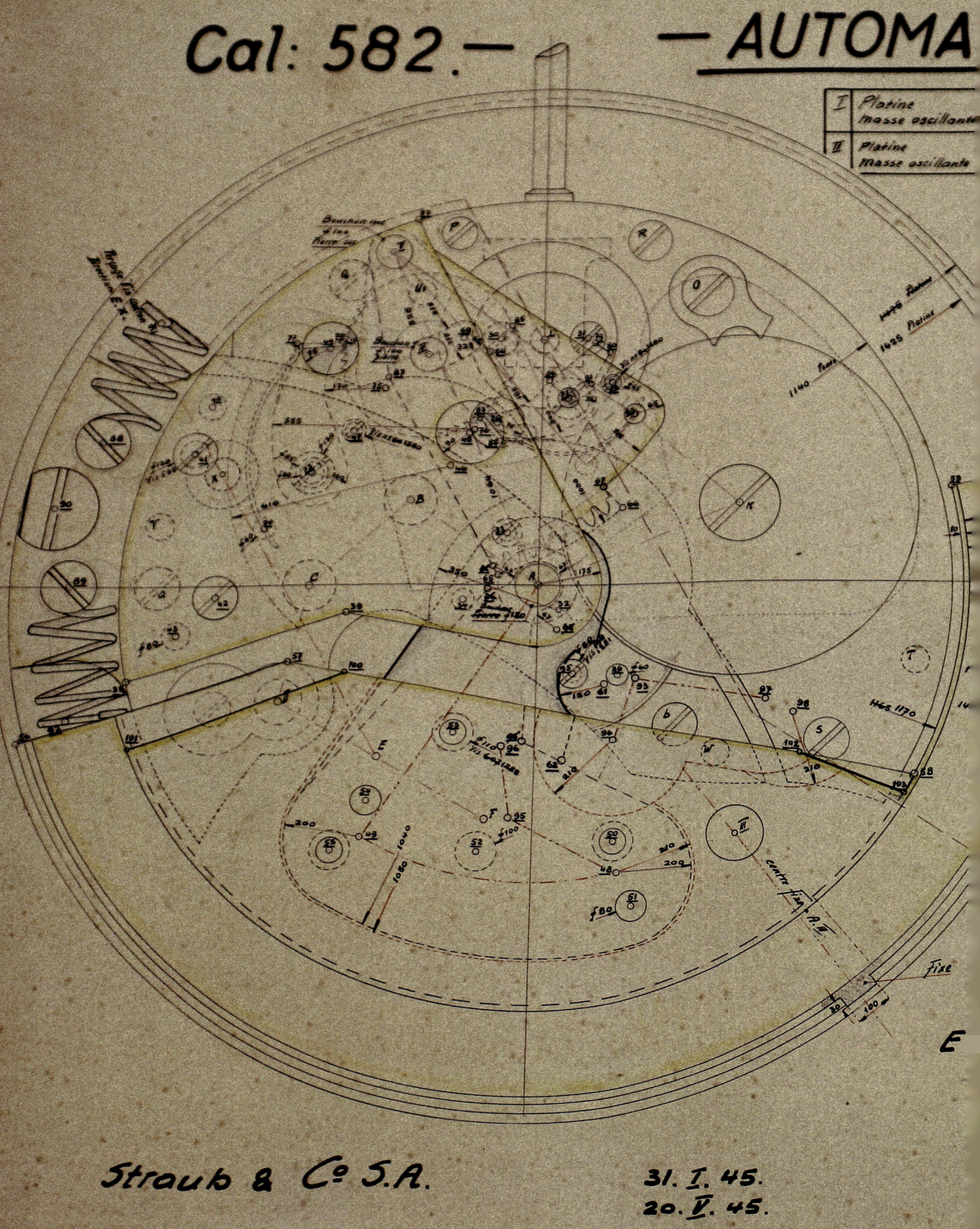
Cal: 582.— —AUTOMA
I Platine masse oscillante
II Platine masse oscillante
Fixe
E
Straub & Cº S.A.
31. I. 45.
20. V. 45.

…IQUE — Cal: 584 — S^de au CENTRE.

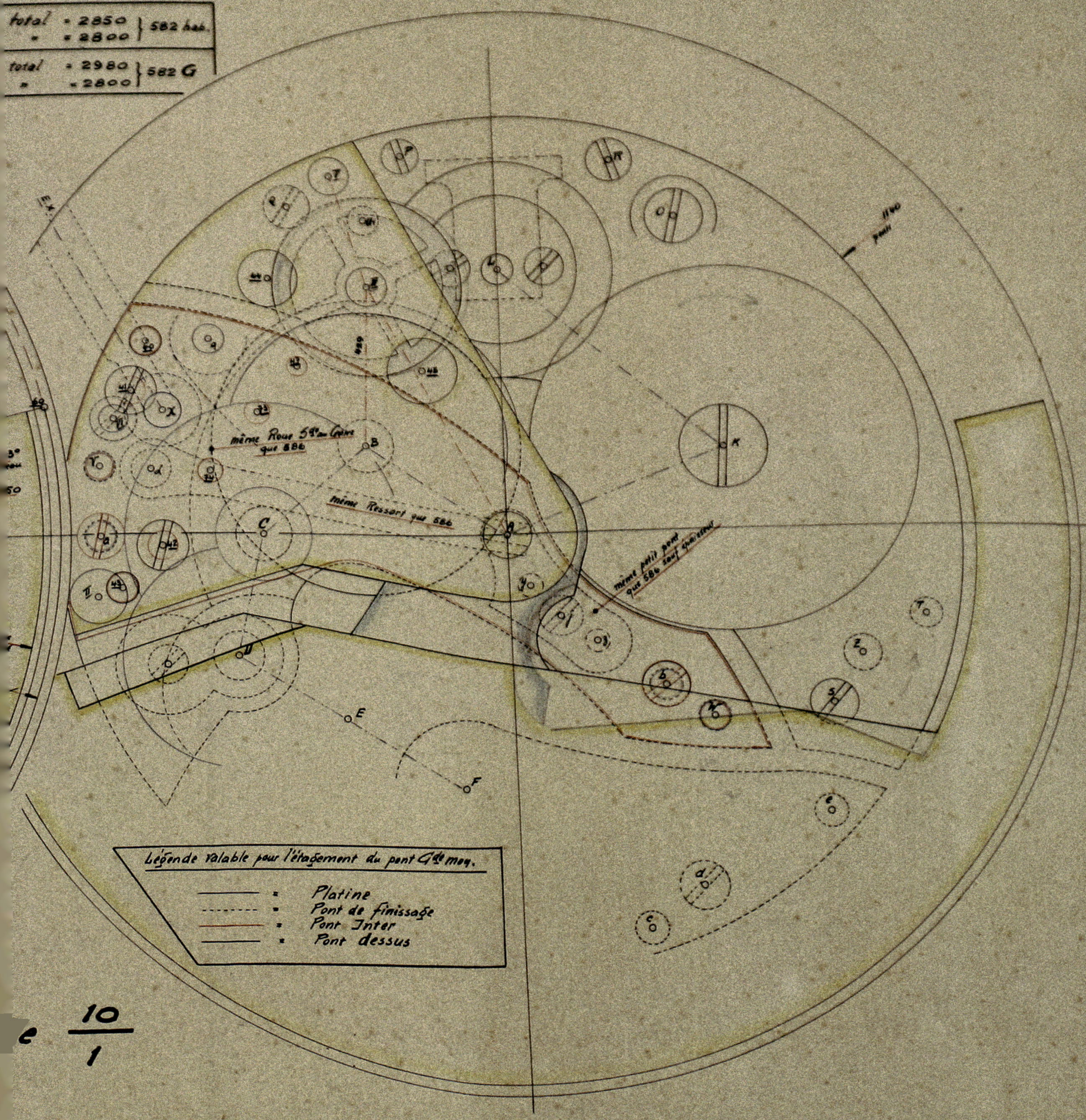

e 10/1

3. X. 47. R. Tschlagen

for the seconds at the "9," also sometimes served as the basis for production in Glashütte. As an expression of especially high-quality craftsmanship, each was equipped with an elegant swan's neck fine adjustment mechanism for its regulator. At the beginning of the 20th century, each caliber boasted an impressive name. Alongside Alpina, the Union Horlogère also used calibers with names like Elite, Eminent, Lilie, Lord, Maxim, Rekord, Roland or Romeo. The enterprise usually identified its products with the hallmarked letters "U.H." or the words "Union-Horlogère." When dealers ordered spare parts, they had to rely on so-called "movement finders" to find the exact appellations. For simplicity's sake, numbers replaced names in the 1930s: e.g. "466" stood for a 19.2 by 12.8 millimeter noncircular hand-wound movement. The tonneau-shaped "475" measured 25.5 by 17.7 millimeters and the rectangular "490" or "Antivibration" was 25.9 by 19.4 millimeters in size. Alpina continued to manufacture gents' pocket-watches in 1938, for which purpose it reserved the 49.6 mm Lépine caliber 355 and the equally sized savonnette caliber 356. These were augmented by the 38.3-mm-diameter Lépine caliber 365 and the 43 mm diameter Lépine caliber 409 or "Magnum," as well as the savonnette caliber 410. All of these calibers were manufactured by Straub in Biel. The era of self-winding watches arrived at Alpina in 1944. Due to patent restrictions, the majority of these self-winding models, all of which upheld Alpina's customarily high quality, relied on a unidirectionally active swinging weight to tighten the mainspring which, when fully wound, contained enough power to keep the watch running for circa forty hours. The 12½-ligne, 5.5-mm-tall, self-winding caliber 582, also known as the "P82," had a subdial for the seconds at the "6"; number 584, also designated as "P84," as well as the 584C with date window, each offered an indirect central second-hand. All of Alpina's other automatic calibers relied on third-party movement-blanks fabricated by experienced ébauche-makers.

In 1948, the school of watchmaking at the cantonal polytechnic in Biel chose 12-ligne hand-wound caliber 592 to serve as the school's didactic caliber. This choice underscores the fact that this device, which had been developed by Straub, was an especially robust, precise and reliable movement.

The 18½-ligne calibers 271, 273, 274 277, 278 and 279 were all based on Unitas ébauches. Moeris delivered the 19-ligne 291 and 291. Cyma-Tavannes provided the 17-ligne 323. Marc Favre in Geneva relied for its calibers on ébauches from Universal Genève for its calibers. Gruen, and thus Aegler, were the sources for calibers 106, 107, 117, 119, 123, 155, 157, 161, 300 (Tecno, constructively identical with Rolex T.S.), 311, 315, 320, 362 (shaped movement, rotor winding), 427 (very-thin), 431, 433, 819 (Rolex 600), 823 (Rolex 500), 827SS (Rolex 700), 833 Rolex 250), 835 (Rolex 200) and 835SS (Rolex 210). Minerva was represented in the caliber palette with its 13-ligne chronograph 20, which Alpina renamed "890." The 19-ligne Minerva 14 was likewise renamed "Alpina 911." The 13-ligne chronograph caliber 940 was in fact the Landeron 13. And the 14-ligne chronographs 942 and 943 were each animated by a Valjoux 22. Also encased were the chronograph ébauches Valjoux 61 (17 lignes) and Valjoux 23 (13 lignes). Other suppliers from the circle of the ébauche giant Ebauches SA were AS, Eta, Felsa, FEF, FHF, Peseux and Venus. Throughout the decades, Alpina's pocket watches and wristwatches encased about 200 different calibers, some of which were exclusive, while others were modified standard items.

BESTSELLERS AND PATENTS

Alpina first began manufacturing sporty wristwatches in 1933. These included the solid steel "Block Watch," which was a highly robust wristwatch with a new crown that had been patented the same year. The innovation consisted of a very close-fitting connection between the crown and the winding stem to prevent dust from penetrating. The manufacturer also claimed that this crown was unlikely to be torn out or lost. Dealers could buy the new product for various calibers in a box of thirty-six so they could quickly supply a replacement whenever servicing was needed. In 1938, i.e. after divorce from Gruen and now doing business solely on its own as a recent divorcée, Alpina presented the "Alpina 4" sport watch. The number "4" was not accidentally chosen; rather, it alluded to four decisive qualitative characteristics: 1. antimagnetic, 2. water-

ADVERTISING
Illustration from the 40's

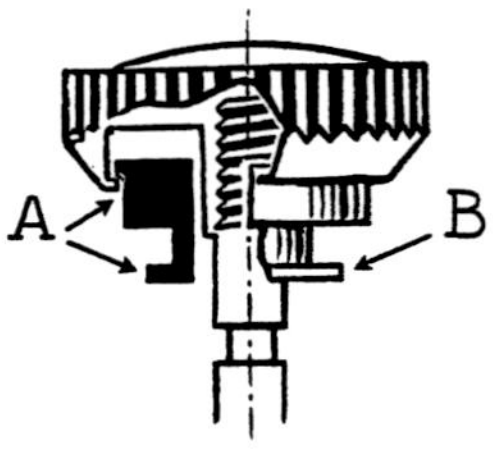

PATENT
Alpina patended crown (1933)

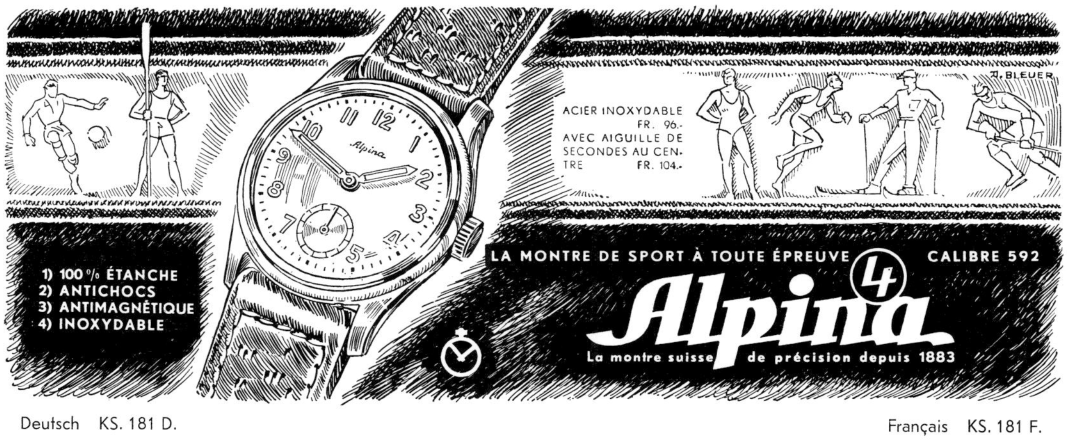

tight "Geneva" case, 3. Incabloc shock absorption and 4. rust-free stainless steel for the highly robust case. Alpina used caliber 566 or 586 for this sporty model, which was available in different versions, for ladies with 8¾-ligne hand-wound movements and for gents with 10½-ligne hand-wound movements. Caliber 586 was offered with either a small subdial for the seconds or with a centrally axial "sweep" seconds hand.

The reliable hand-wound caliber 592 served faithfully for several decades in successful sport watches such as the Alpina 70 (which debuted in 1953) the Standard (1958) or the Tropicproof (1968). When the new Alpina President debuted in 1957, it gave automatic caliber 584C a suitable stage on which to show off its abilities. Naturally, discriminating ladies were unwilling to make do without self-winding watches of their own, so Alpina took appropriate action in 1963. For shaped caliber 362 with rotor winding, Alpina revived its traditional relationship with Gruen in Biel, where the thoroughly unconventional caliber 362, which measured 15.3 by 17.2 millimeters, had become ready for serial production around 1955. In its best decades, Alpina's collection included several hundred references. This was impressively evident at the annual congress in 1958, when Alpinists from around the world could admire some 1000 different models. As in previous decades, here again the collection clearly focused on sporty wristwatches.

ANOTHER DIFFICULT ERA: WORLD WAR TWO

To assure its survival during the First World War, the Union Horlogère had been obliged to split into three legally independent companies. Because of its international activities, the firm and its business relationships were again subject to close scrutiny after the outbreak of the World War Two. The restrictions affected imports, travel and the flow of capital, thus significantly reducing the scope of potential activities. The Allies' constraints culminated in prohibiting the Swiss Alpina Union Horlogère from using the Alpina brand name in Germany. Afterwards, the German association had no other option but to adopt a new name (Deutsche Uhrmacher-Genossenschaft G.m.b.H) and to use this as the insignia on its watches. The end of the war saw an intensification of collaboration between Alpina and Dugena. The Union Horlogère delivered Alpina-brand Swiss timepieces to Dugena. . Dugena also continued to manufacture their own brand of watches. Financial irregularities shook Dugena in the 1970s. In the wake of the takeover of a supermarket chain, the firm was converted into a joint-stock public company in 1984. Dugena was sold to the German Christ-Holding in 1988. Egana-Holding, with headquarters in Hong Kong, , came onboard in 1993, but the new owner suffered a severe financial downturn in 2007. A new owner was once again found Nova Tempora Uhren and Schmuck GmbH in Meisenheim, Germany.

The history of Union Horlogère AG in Biel is a similarly turbulent and at the same time a colorful narrative. After the end of the Second World War, the business initially grew and developed very well. But as is so often the case, too much success can tempt a brand to rest on its laurels. Good revenues blinded the decision makers, who failed to anticipate the oncoming Quartz Revolution. By the early 1970s, Alpina found itself helplessly confronting the market's irresistable new trends. The Union Horlogère had little to offer in reply. For too long, the Straub Family, as majority shareholders, had relied on the strengths of their affiliation with the Alpinists as loyal and dedicated retail dealers. A name change and new proprietors were unavoidable in 1972. The majority of shares in the new Alpina Watch International AG/SA in Biel were now owned by the Gerl Brothers in Cologne. At the same time, Gerl, Dohrmann & Layer took over Alpina's general agency for Germany. Names such as President, La Belle, Comtesse and Seastrong were added to the list of the brand's traditional product lines. Another revision by the Gerl, Dohrmann & Layer Group ensued in 1977, when investors placed Alpina under the aegis of Montana Uhren AG in Germany. From this moment on, the business's motto became "Alpina – The Brand for Selected Retailers." The business's formerly clearly defined marketing strategy became increasingly vague, and the "Alpinist Spirit," which had historically made Alpina so strong, was gradually lost. The brand lacked the clarity of a well structured and consistently styled watch collection. Its spectrum consisted for the most part of "me too," copycat and random products. After the devastating effects of the Quartz Wave, Alpina then subsequently slept through the renaissance of mechanical watches, which began in the mid 1980s, perhaps because the German owners of Alpina Watch International SA were both geographically and commercially too far removed from the Swiss epicenter of ticking events and the remaining Alpinists. Year after year, the traditional brand drifted more or less without direction. Towards the end, it had one foot in the grave and seemed inexorably on its way toward the archives of horological history. Fortunately, the Dutch entrepreneur Peter Stas came onto the scene at the eleventh hour and discovered the brand's slumbering potential.

CHRONOGRAPH 482
Alpina Chronograph from the 30's

HEADQUARTERS
Alpina Manufacturer in Biel

STARTIMER
Relaunch New Collection
2003

That's the long, often exciting and quite colorful story of Alpina, Peter. Would you like to add anything to the narrative?

PETER: Our distributors repeatedly asked us to launch a line of sport watches, but we categorically rejected their requests. Frédérique Constant is not a brand with sport watches and it never will be. Of course, people told us that other brands offer both genres, but we were strictly opposed to the idea because we didn't want to detract from the clarity of Frédérique Constant. Sport watches wouldn't have been appropriate alongside the classical models in our little displays in shop windows.

So a brand like Alpina was exactly what you were looking for?

Yes, that's right. The traditional brand had spent seventeen years in the hands of the Gerl Brothers in Cologne, where the label slowly but surely atrophied. Only thirty shops remained in Germany. Production was very small, the watches looked awful and they sold for between 200 and 300 deutschmarks. The owners tried everything to revive the brand, but they never succeeded. There were no design ideas. Everything was controlled from Cologne. There still remained an administrator, a bookkeeper and a lawyer in Biel who were paid each year by the brothers in Cologne. Needless to say, the business was swimming in red ink.

What happened next?

I drew a few design sketches and visited Cologne several times to explain my strategies for the brand's future. My ideas were well received. That created the necessary connections for more specific negotiations, and finally we signed a contract for the firm's immediate takeover. The first installment of the agreed purchase price was payable after three years and the final installment was due after six years.

How did you begin talking business with Alpina?

Our first meeting was at the Inhorgenta watch fair in Munich. Our agent in Germany knew the two brothers and it was he who initiated our first contact.

In your view, what were Alpina's principal values?

The history of the Alpina watch brand is much older and longer than that of Frédérique Constant. Alpina is Swiss through and through. It has a short and catchy name that's suitable for sport watches and offers much more stylistic latitude than Frédérique Constant. We would be able to make watches under this label that would be more "German," i.e. larger, bolder and more angular.

When did you first come onto the market?

Our first collection was ready in 2003. We initially thought that it would run as smoothly as Frédérique Constant, but we were completely mistaken. Alpina meant starting again at zero, just as we had with Frédérique Constant in the early 1990s.

Are you satisfied with the evolution?

Annual production has now reached about 8,000 watches. We're aiming for annual totals between 25,000 and 35,000 watches in the midterm. That's how many wristwatches will be necessary to make the brand better known and to generate developmental potential for new items. We haven't yet reached our goals with Alpina. When I compare Alpina's first decade with Frédérique Constant's first ten years, I would definitely say that we worked better, in principle, with our original brand. We hadn't sold as many watches in Alpina's current price class, but we had developed more favorably. Alpina's first ten years were marked by very hard work.

Did you ever regret the purchase?

Not for a minute. The brand never lost money under Frédérique Constant's aegis. It's like a baby within a larger business and now it simply needs to grow up. For this purpose, we've employed personnel expressly for Alpina, which hadn't had its own designated staff in the past, when Alpina had been allowed, so to speak, to run alongside Frédérique Constant. But that's not enough. And – to be perfectly honest – we hadn't invested enough either. After the change in 2012, it's going much better. And we appointed a commercial director for Alpina at the beginning of 2012.

Alpina is now at home in the air, on land and in the water. Where does the brand feel most comfortable?

Our pilot's watches have been very well received by customers since 2011. That's why we're planning to further expand this theme in 2013. We also had diver's watches in the past and they're developing very satisfactorily in 2012. If we had done more for them, things would have run even better. Several things will happen here next year. But the two most important pillars, i.e. pilot's watches and diver's watches, offer no cause for fundamental criticism. We have our rally watches as our third pillar. And the fourth pillar is the "Heritage." We made a promising start with the "Heritage Pilot at Baselworld in 2012. It's a model that we can definitely build upon.

AL-725LBS4S6B GMT Chronograph

AL-525LB4S6 GMT Automatic

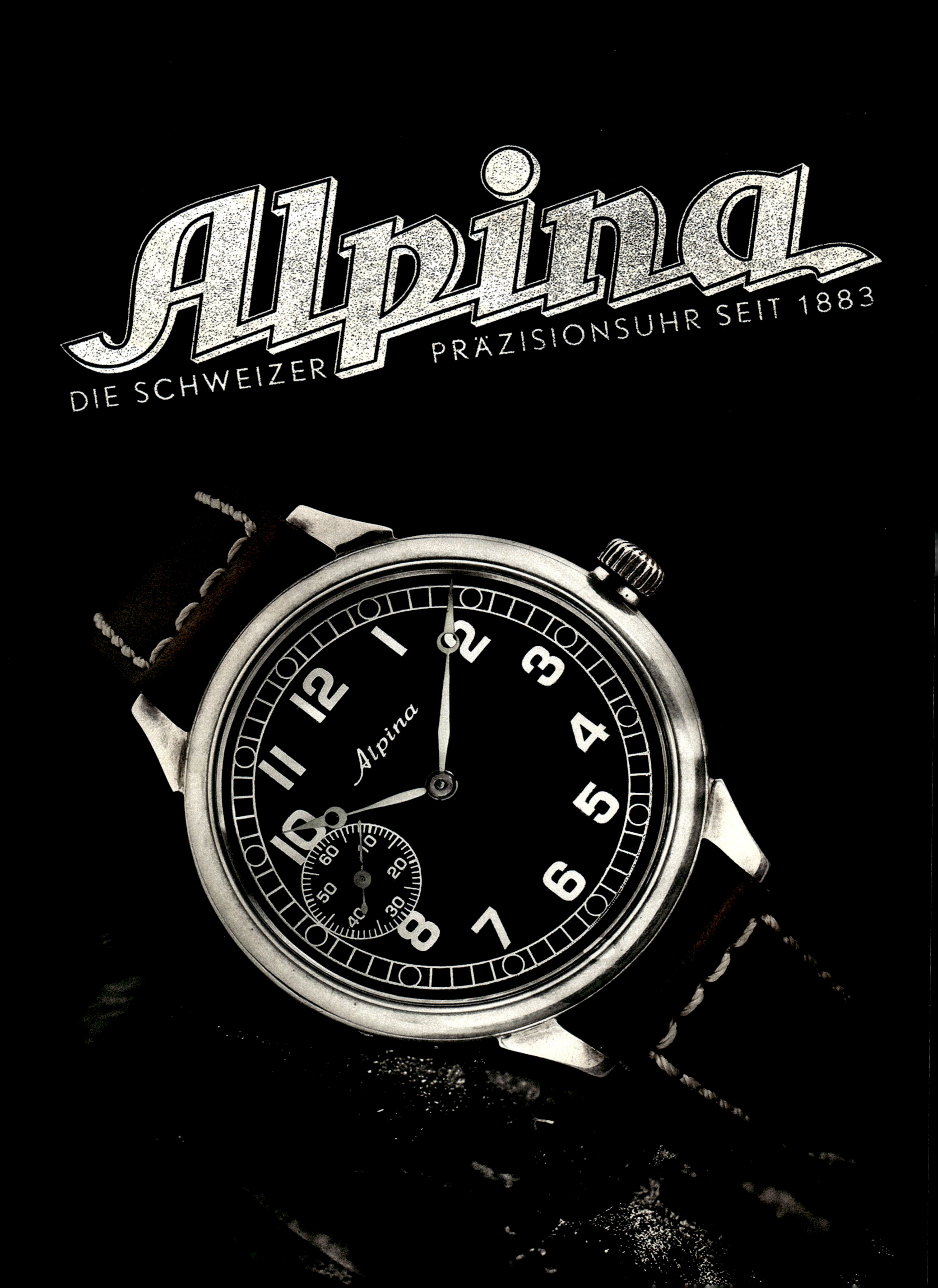
Alpina
DIE SCHWEIZER PRÄZISIONSUHR SEIT 1883
Alpina

AVIATION AND FINE WATCHES:
A PERFECT PARTNERSHIP

December 17, 1903: Orville Wright lies prone upon the wing of the Wright Flyer, preparing for takeoff. The homemade engine of the little airplane pops and snarls as it spins the two eight-foot, hand-carved propellers into opaque disks. The cold, blustery wind sweeping in from the sea chills the pilot and rocks the wings of the pioneer plane. The primitive airplane built of wood and cotton fabric is about to carry Orville Wright and all of mankind into the air age.

Cessna
N1098Q

AVIATION HISTORY
First flight Wright Brothers 1903. Cessna from 1927.

CHRONOGRAPH 943
Alpina Chronograph
from the 40's

Only three instruments were available to Orville on that very first powered, heavier-than-air flight: a tachometer, an anemometer and a watch. And as the watch faithfully measured, this momentous first flight lasted exactly 12 seconds.

From the very beginning, aviators' lives depended upon the ability to accurately track time while in flight. Pioneering aviators blazing new trails across oceans and continents relied upon a primitive form of navigation that balanced estimated speed and direction of flight, against the accurate measurement of time. At the end of a long flight, those early aviators emerged from their open cockpits wind-tousled, oil-spattered, and wearing rumpled, grease-stained flying togs. But they often sported the most expensive of fine timepieces on their wrists or in their pockets, valuing the precision accuracy and reliability afforded to them by these luxury timepieces.

John Alcock, who along with his partner Arthur Whitten Brown was the first to fly the Atlantic nonstop in 1919, died in a crash just months after his famous flight. He was found in the wreckage of his plane wearing a diamond-studded wristwatch. When Amelia Earhart made her record setting flight as the first female pilot to fly the Atlantic in 1932, she was wearing a two-register chronograph wristwatch. That very same watch was carried to the International Space Station in 2010, and is now on display at the Ninety-Nines Museum of Women Pilots.

Jimmy Doolittle is considered by many aviation historians to have been the greatest of the aviation pioneers. History buffs know of Doolittle as the leader of Doolittle's Raiders, the bomber squadron that took off from the deck of an aircraft carrier in the first American strike against the Japanese mainland in 1942, as dramatized in the Hollywood movie Thirty Seconds Over Tokyo.

However one of Doolittle's greatest achievements occurred years earlier, in 1929, when he made the first flight in history to be flown solely by instruments. Doolittle took off, navigated a course, and landed without ever seeing the ground, completely reliant upon his precision flight instruments and, of course, an accurate timepiece. One of Doolittle's wristwatches, a gold-plated watch that incorporated three miniature compasses in the dial, is now part of a collection at the National Air and Space Museum in Washington D.C.

Even as the air age advanced into the space age, "the clock is running" post-launch report to mission control by the astronaut in command confirmed the functioning of that all-important instrument, the timepiece.

GOLDEN AGE AVIATORS DEPENDED UPON ALPINA WATCHES

The 1920's and 1930's were considered the Golden Age of aviation. Pilots like Alcock and Brown, Earhart, and Doolittle achieved worldwide fame for their aeronautical achievements, setting records for speed, distance and altitude on a regular basis. Nomadic pilots known as barnstormers carried the excitement of aviation to small towns across the land, sky-hopping thrill seeking passengers from cow pastures and farm fields.

Many pilots of the time relied upon Alpina watches. Several Alpina models in the 1920's and 1930's were highly popular among pilots. And Alpina watches were also held in particulary high regard by military aviators, with Alpina designated as the supplier of watches for the air forces of several nations during this time.

STARTIMER PILOT
New Alpina Startimer Pilot
Collection 2011
AL-525B4S6 Automatic
AL-860B4S6 Chronograph Automatic
AL-710B4S6 Manufacture Automatic

Alpina
GENEVE
Alpina
GENEVE
AUTOMATIC
CHRONOGRAPH
Alpina
GENEVE

PIONEERING PARTNERS

Alpina, founded 20 years before the Wright Brother's first flight, was coming into its own as a world-class watchmaker when another famous company was getting its start. Cessna Aircraft Company, founded in 1927 by Clyde Cessna and Victor Roos, grew to become the largest aircraft producer on the planet.

Since its inception, Cessna has produced more than 190,000 aircraft, more than any other manufacturer in the world. One Cessna model alone, the 172 Skyhawk, has been produced in greater numbers than any other aircraft model.

Just like Alpina, Cessna continues to be a cutting-edge company and a leader in its field. Cessna's Citation line of business jets has been in production for more than 40 years, and remains the world's best-selling line of business jets. And Cessna's new Skycatcher light sport plane is a state-of-the-art leader in the field of recreational aviation.

It's fitting, then, that Alpina and Cessna have teamed up as marketing partners, each delivering superior, elegant, cutting-edge designs to the aviation community.

A Tradition Continues

With its new Startimer collection, Alpina continues its tradition of supplying fine watches to the world's aviators. The Startimer line offers pilots the accuracy and dependability they require in the cockpit, along with the sophistication and elegance that are so valued outside of the cockpit. Models such as the automatic chronograph AL-860 and the Manufacture AL-710 combine the clean, classic beauty of Golden Age watch designs with modern, precision engineering and manufacturing.

STARTIMER PILOT
AL-280B4S6 Alpina Startimer
Pilot Big Date (2012)

Alpina
GENEVE
SWISS MADE

Alpina
SWISS
MADE

ALPINA AVIATION

The beauty of flying is infinite; it mixes the feelings of fear and admiration. For a pilot, each flight is a breathtaking experience. When you are thousands of feet up in the air, the clouds are beneath you and everything is calm, but you know, in the back of your mind, that you are at nature's mercy and you have to keep your eyes constantly on the controls. Your muscles relax and you enjoy the flight.

In the early twentieth century, Alpina was known as a specialized manufacturer and official supplier of military watches, almost exclusively for pilots. These early aviation pioneers had strict requirements. Their instruments had to be extremely precise and provide ultimate shock resistance. In addition, Alpina pilot's watches were mostly equipped with anti-magnetic cases, high-contrast dials with oversized luminous numerals and luminous hands for instant readability.

In 2011, almost a hundred years later, Alpina went back to its rich aviation heritage and introduced the new Startimer Pilot Collection, offering four exciting variations, inspired by original Alpina designs from the 1920s and 30s. These included a basic three-hand date automatic (cal. AL-525), a timepiece with in-house developed and produced automatic caliber AL-710, featuring minute and hour indications and off-center date hand, an in-house caliber AL-950 that powers the iconic Startimer Pilot Regulator, combining historic pilot watch design features with a modern day regulator lay out and, last but not least, a well balanced and powerfully designed automatic chronograph version (cal. AL-860).

HERITAGE PILOT
AL-435B4SH6 Reedition of an original design from the 20's

ALPINA • GENEVE •
SEVENTEEN
17 JEWELS
SWISS
Alpina
Limited Edition

The clear design that connects these timepieces was clearly inspired by Alpina's past. Each watch boasts key features such as generous 44 mm, anti-reflective satin-finished stainless steel case, matt black dial with oversized white luminous numerals, a glare-free sapphire crystal and a signature red Alpina triangle on the seconds hand. To further improve user-friendliness and comfort, a clever solution was found to secure the leather strap. It closes on the front side of the wrist, not on the reverse or lower sides, thereby resembling the vintage leather straps of the past.

To support the launch of this important new collection, Alpina proudly joined forces with Cessna Aviation and PrivatAir, an airplane manufacturer and a private jet airline, signing a three-year agreement with both partners. During three years, Alpina, Cessna Aviation and PrivatAir work together closely to promote their equally high quality products and services, through cross-marketing activities in Europe and North America

We both rely on precision, hand-crafted products to please our customers

Furthermore, the production of the Alpina Startimer Pilot timepieces is presented in a dedicated, luxurious packaging, alongside a Cessna Citation Mustang scale model in PrivatAir livery. To celebrate the first deliveries of the new Alpina Startimer Pilot Collection, Alpina organized an exciting event in collaboration with Cessna Aircraft and PrivatAir. After a joint presentation of Alpina Genève, PrivatAir and Cessna during which journalists were able to rediscover the Alpina Startimer collection and to understand the motivations of this great collaboration, the members of the press were invited to board Cessna Citation Mustang & Cessna Citation Excel private jets for a memorable flight from Geneva to Chambéry airport (France), by making an incredible tour over the Alps at very low altitude. The summits of the Mont Blanc and Matterhorn were almost in touching distance and the view of this magnificent landscape was just amazing.

HERITAGE PILOT
Alpina Heritage Pilot caseback – Handwind movement visible on the hunter case

370

SAPPHIRE - AL280X4SP6 - ALPINA - STARTIMER PILOT - SWISS MADE - 10 ATM
0013/8888
LTD EDITION

CAUTION

PRESS TOURS
Visit of the CESSNA Plane Manufactu re in Wichita – Kansas
Flights in Switzerland and New York

Alpina
GENEVE

The flight was followed by a exquisite lunch above the 'Lac du Bourget' in Chambéry, during which the Alpina Startimer Pilot timepieces were shown and discussed in-depth. We showed the AL-710 Manufacture Startimer Pilot, built with an in-house developed and produced automatic caliber that features minute and hour indications as well as off-center date hand.

The movement was designed and developed entirely in the manufacture's workshops in Plan-les-Ouates, Geneva, and features an automatic winding mechanism incorporating Alpina's trademark rotor – a salute to Straub & Co., the designers of Alpina's first automatic calibre , cal. 582, from 1945. After considerable investments in technical know-how and, most importantly, a highly motivated team of experts, Alpina's ambition of manufacturing a new bespoke in-house calibre was accomplished. As with all Alpina mechanical timepieces, the finely decorated movement is visible through a transparent case back.

In August 2011, Alpina and Cessna organized another exciting event at Cessna Aircraft's headquarters in Wichita, Kansas. Selected members of the press and VIP customers attended a presentation and viewed over 30 airplanes in different stages of production. Thereafter various Cessna planes were inspected in the main showroom at Cessna, including a brand new Cessna Citation Mustang. Since Cessna and Alpina are essentially talking to the same people, cooperating through events like this benefits both companies.

STARTIMER PILOT 18K GOLD
AL-710B4S9 Alpina Startimer Pilot in 18K rose gold with Manufacture caliber Limited edition of 188 pieces

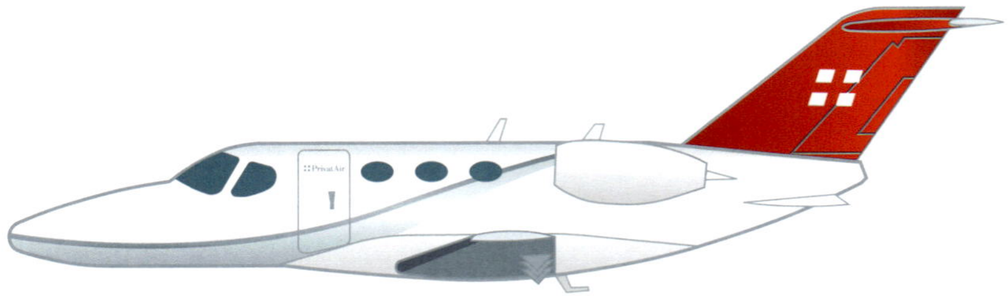

MODEL PLANE
CESSNA Mustang Private Jet in PrivatAir Livery

SPECIAL PACKAGING
Alpina gift box for the Startimer Pilot watch models

MORE ABOUT CESSNA AIRCRAFT

This renowned airplane manufacturing company headquartered in Wichita, Kansas, is a subsidiary of the US conglomerate Textron, a company operating in 25 countries. Since its creation in 1927 Cessna's strong devotion to high quality workmanship, precision and creativity and its passion for the aircraft industry, have led it to be an inspirational leader. Cessna is now the leading manufacturer and inspired designer of innovative, light and midsize business jets, utility turboprops and single engine aircraft.

The Cessna Citation Mustang is an entirely new kind of business jet, one that's designed - and priced - to bring the dream of jet ownership comfortably within reach of a whole new class of upwardly eager operators. It features ample cabin class seating, true jet speed and a good half-continent range. Extraordinarily high levels of performance and assurance are now available in a jet aircraft costing much less.

MORE ABOUT PRIVATAIR

PrivatAir is a leading international aviation group with headquarters in Geneva, Switzerland and operating bases in Zurich, Düsseldorf, Frankfurt, Munich, Amsterdam and Bahrain. Services include Private Charter for VIPs and executives, Aircraft Sales & Management, Private Airline Services, Ground Services and Training. PrivatAir gives a luxurious service at an accessible price.

PrivatAir

Alpina
GENEVE
AUTOMATIC
300 M
SWISS MADE
Geneve

ALPINA EXTREME DIVER COLLECTION

Time is a very crucial thing in diving, because you only have a certain amount of time before you run out of oxygen, and a broken or inaccurate watch can be fatal. Because your safety is an absolute priority, Alpina builds the most robust diver watches which can go as deep as 300 meters.

There is a kind of beauty to the professional diver's watch that comes from the pure pleasure of engineering excellence. To better appreciate the Alpina Extreme Diver, the first thing to do is understand the requirements that define the professional diver's watch. In this regard, it is possible to define clear parameters and, without resorting to hyperbole, make the case for the true professional diver's watch. As a general rule, for anything from snorkeling to scuba diving, a diver should always choose a watch with a water resistance rating of at least 300 meters. This applies if these are activities a diver will be regularly subjecting a watch to. For the occasional swim, a watch resistant to 100 meters will do. You might think that 300 meter deep water resistance is a lot, but the rationale behind this is that, in general, watches tend to perform differently in real-world situations than they do in the laboratories where they receive their certifications and you don't want your watch to break when you are diving.

EXTREME DIVER
AL-525LB4V26 Alpina Diver Automatic (2011)

6883525 - AL525X4V26 - ALPINA
30ATM
Alpina
GENEVE

The case of the Alpina Extreme Diver is made of anti-corrosive materials has non-scratch crystals, and is outfitted with extendable wrist straps that can fit over a wet suit. The bezel of the Alpina Extreme Diver is unidirectional and has luminous markers. The bezel of the Alpina diving watch is used to set the maximum time a diver plans to be under water so it is vital that it cannot be accidentally pushed or knocked off its setting. Having it move in only one direction means that time meant to be spent under can only be accidentally reduced, not increased.

To support the introduction of it's new Extreme Diver collection, Alpina contracted William Baldwin as the new face of its 2012 advertising campaign unveiling the latest model in its series of Diver Watches. In June 2012, we were pleased to be in New York City to celebrate the new Alpina Extreme Diver Collection with Billy Baldwin. The Alpina event took place at The Maritime Hotel North Cabana with an announcement to officially introducing William Baldwin as the brand ambassador, as well as announcing a new partnership with Baldwin's family charity for breast cancer research: The Carol M. Baldwin Breast Cancer Research Fund.

EXTREME DIVER
AL-525LB4V26 Alpina Diver
Automatic

William Baldwin
Alpina Brand Ambassador

Alpina
WATCHES

"We were looking for a famous face to support the introduction of our Extreme Diver Collection. With Billy Baldwin, we found a supreme celebrity. Billy is cool and courageous, with the same values of the Alpina brand"

Alpina made a donation of ten Alpina Extreme Diver Watches to The Carol M. Baldwin Research Fund to be used as auction pieces for their various galas and events, which are held to raise money for breast cancer research. These special edition Diver watches were produced in a limited quantity of 10 pieces, with an engraved silver plate enclosed in the box featuring The Carol M. Baldwin Breast Cancer Research Fund logo. In addition, the case back of the watch also shows the same delicate engraving in tribute to the Carol M. Baldwin Research Fund.

"It is truly an honor to work with a brand like Alpina," *said Baldwin.*

"They not only represent everything that is bold and innovative, but also that which is caring and giving. I look forward to a successful relationship with them."

Baldwin describes the watch as follows:
"The new Alpina Extreme Diver is that ideal accessory which combines the elements of diving while delivering a bold, athletic, cool-factor that's completely unrivalled."

page 242-243
ALPINA DIVING
In action (lt), Extreme Diver AL-525LBV26 (lb), William Baldwin, Alpina Ambassador (r)

Alpina
AUTOMATIC
300 M

Alpina
WATCHES

Alpina
300m/1000ft
AUTOMATIC
SWISS
MADE

ALPINA SAILING

EXTREME SAILING
AL-880LB4V6 Automatic regatta caliber with countdown chronograph (2011)

GET READY TO TACK!

shouts the Tactician. Everybody prepares themselves, blood is pumping faster, the grinders are on their positions and the ship starts turning. "Tacking!" The sails switch sides with a loud rumbling noise. Everything has to be fast and precise. No mistake can be made. The 18 man crew has to act as one coordinated body. As the wind blows through your ears and the water splashes on deck, you concentrate on your task to win this regatta.

In 2011, to promote the introduction of its Sailing Collection, Alpina sponsored Team Niceforyou, sailing regattas around the world in the Extreme Sailing Series. Niceforyou believes in the quality and high standards of the Extreme Sailing Series and decided to take part in the competition from 2011, alongside a team of extremely high profile sailors, led by skipper: Alberto Barovier. Barovier was the winner of the 2010 America's Cup on board the trimaran BMW Oracle Racing. The team gathers some of the best sailors from Italy, who together boast 12 participations in the America's Cup, more than 15 world titles, a similar number of European and national titles and one Olympic Medal.
The development of the Alpina Sailing Collection was inspired by the close collaboration with this Italian Extreme 40 Catamaran team Niceforyou. Each Alpina Sailing timepiece is offered in limited edition and is delivered in a specially designed package alongside a miniature reproduction of an Extreme 40 Catamaran.

page 246-247
ALPINA SAILING
Alpina is sponsor of the Exterme Sailing Series Team Nice-for-You (2011), AL-880LB4V6 (mb)

Alpina
GENEVE
ALBI BAROVIER
Alpina
GENEVE
TOE IN THE WATER
EXTREME
SAILING SERIES
PETERS & MAY
RACING
GLOBAL BOAT TRANSPORT
Niceforyou
Alpina
WATCHES

EXTREME
SAILING SERIES
niceforyou
.com
TOE
Alpina
WATCHES

TOE IN THE WATER
COMPAGNIA DELLA VELA

Alpina
GENEVE
300m/1000ft
AUTOMATIC
SWISS MADE

GURNARD
Alpina
GENEVE
Alpina
Niceforyou

Alpina
GENEVE
300m/1000ft
AUTOMATIC
SWISS
MADE

Its star product is an automatic chronograph with regatta start function. Alpina developed this regatta caliber AL-880 in-house, based on the Sellita SW500 chronograph movement. The caliber is encapsulated in a stainless steel 30 ATM water-resistant case with rotating bezel. The cases shape resembles Alpina case designs from the past. Currently, Alpina is the only watchmaker offering such an easy to read automatic regatta caliber.

This model with sailing countdown function is available either with a white or a black dial. In addition to the regatta countdown version, an attractively priced Alpina Sailing Automatic with three hands and a date window was introduced, with similar case and dial design. All these Alpina Sailing timepieces are available with rubber strap, metal bracelet or metal mesh bracelet, for comfort and security in the most extreme situations.

In 2012, to continue support for the Alpina Sailing Collection, we announced the branding of a Swan 82 which was participating in various yacht regattas in the Mediterranean. The Alpina Swan raced in the Palmavella, Loro Piani, Giraglia and Voiles de St. Tropez regattas.

SAILING ADVENTURE
AL-525LB4V6B

Artemis
Racing
Fly
Emirates
DEAN BARKER
NESPRESSO
E IN THE WATER
USCAT
NESPRESSO

EXTREME SAILING
Alpina partnership with
the Extreme 40's team "Nice for you"

120
100
80
60
40

EXTREME SAILING
AL-880LB4V6B Automatic
regatta caliber with countdown
chronograph (2011)

Why is Alpina involved in sailing regattas?

PETER: We support the Alpina Sailing Collection through our sponsorships of various regattas. The star product of the Alpina Sailing Collection is an automatic chronograph with regatta start function. It has a unique and easy to read regatta display. In the past, a Valjoux caliber offered the same layout. ETA discontinued it and we decided to develop a module in-house for this regatta function. Alpina is now the only watch manufacturer offering such an automatic regatta caliber. To promote this watch, we have started limited advertising. The most important promotion however is to bring this watch to sailing enthusiasts during regattas.

You are on the helm of the Alpina Swan, what is driving you?

PETER: I have sailed in regattas since I was a child, as a sailing instructor and for leisure. Starting in small boats in The Netherlands, I later sailed larger boats in the Atlantic, the Mediterranean and in Hong Kong. Being on the water is a passion and participating in regattas is extremely exciting. It demands ultimate concentration. The teamwork during a regatta is a learning experience that doubles as management training. During a regatta, there is no margin for error, lives are at stake. The team on a boat must run like clockwork.

ALPINA SAILING
Alpina Swan 82 in action during the Palmavella Regatta (2012)

www.alpina

Alpina
GENEVE
www.alpina-watches.com

EXTREME SAILING AL-880LS4V6B Automatic regatta caliber with countdown chronograph (2011)

Alpina
GENEVE
www.alpina-watches.com

Alpina
GENEVE
REGULATOR
SWISS MADE

ALPINA AND AURÉLIEN DUCROZ

(1)

FULLBLACK REGULATOR
AL-650LBBB5FBAE6 Alpina Avalanche Extreme Regulator – Iconic timepiece from the Alpina collection

High peaks, snow-covered mountain slopes, that is the true dream of any skier. You look down and you see more than 1000 metres of descent with fresh, untouched powder. Your feet start to tickle because you know that it's going to be a moment of full enjoyment.

That almost seems like a dream, doesn't it? Well, Aurélien Ducroz does it as his job. Always skiing and getting paid for it; it's almost impossible to find a better job. (1)

ALPINA AMBASSADOR
Aurélien Ducroz – Freeride Ski Worldchampion 2009 & 2011

Born in Chamonix in 1982, son of a mountain guide and a skiing instructor. For Aurélien Ducroz, skiing was the most obvious thing to do. He already started with alpine skiing competition at a very young age and then turns to ski jumping, where he won the France junior championship two times.

Having had enough of tracks and prepared ski slopes, Aurélien changed to high mountain snowboard. He progressed well and managed to enter the high level circuit. However, it was only when he started doing free ride skiing that his true talent got revealed. Aurélien improved continously and won the world freeride championships both in 2009 and in 2011. He is now regarded as the best freeride skier in the world.

Aurélien Ducroz' creative and fluid skiing style has given him lots of victories. He is never afraid, and is always ready to jump from higher cliffs and ski down steeper slopes. Even though he is at the top of the world of freeride skiing, he has always stayed modest and down to earth, a characteristic that cannot be found in all professional sportsmen.

Alpina
GENEVE
NISSAN

BETWEEN
SNOW AND SEA

As if skiing wasn't enough for Aurélien, he also started professional sailing in the summer of 2011 and participated in the Transat 6.50', a single handed transatlantic regatta. One week after his great third place at the Verbier Xtreme in 2012, he was already on the water around the Lofoten islands, North of Norway. He was there to sail and to shoot sequences of a freeride movie with the famous American Producer Waren Miller Entertainment. Aurélien Ducroz wanted to add a bit of his own spice to the movie and combined sailing and skiing. That is why they chose the Lofoten islands to shoot the movie. It is Aurélien's dreamplace; you can ski and sail, on the same day!

"The Lofoten Islands, Norway were part of my projects since a long time, but it was necessary to wait for ideal conditions and in particular to find a suitable boat to sail in these areas to achieve beautiful colors, and to find intouched, lovely skied landscapes: the skiing was incredible. We managed an average of 1500m of difference in height per day (ski touring obviously), but the weather conditions did not allow us to do as many routes as we planned."

Aurélien's constant search for adventure makes him the perfect ambassador for the Alpina Adventure collection. Always wearing an Alpina Manufacture Regulator, it is the perfect way to test its resistance. From ice cold blizzards on high mountains to huges waves and storms on the Atlantic Ocean, the watch didn't break down and Aurélien still wears it proudly around his wrist.

AURELIEN DUCROZ
During the Translatlantic Regatta (t),
North of Norway (b)

AURELIEN DUCROZ Lofoten Islands, Norway

Alpina
GENEVE
12 Hours of
SEBRING
SWISS MADE

ALPINA RACING

Roaring engines, tightened muscles, hands gripped around the steering wheel, pushing the limits to get to the top. After hours of racing, light is fading, the energy is starting to lag, but you stay focused, and you fight the exhaustion. It's all about skill, courage and persistence.

THE 12 HOURS OF SEBRING

The 12 Hours of Sebring is a motorsport endurance race held annually on the Sebring International Raceway, a former Army Air Force base in Florida. The 12 Hours of Sebring is the opening race of the FIA World Endurance Championship and is a leg of the Triple Crown of Endurance Racing, the three most famous endurance races in the world: the two others being the 24 Hours of Le Mans and the 24 Hours of Daytona. The first 12 Hours of Sebring to ever take place was actually the 6 Hours of Sebring, held on New Year's Eve. Fourteen months later, the race time changed, becoming twice as long and the 12 Hours of Sebring as we know it now was born. The popularity of the race kept growing and the 12 Hours of Sebring became one of the classic American car races.

ALPINA RACING
AL-750LBR5FBAR6
Alpina "12 Hours of Sebring"
Chrono GMT Automatic
Limited edition of 158 pieces

page 272-273
ALPINA RACING
Alpina is Official Timekeeper of 12 Hours of Sebring (t, rb), Romain Brandella – Alpina Racing Ambassador (lb, mb)

OFFICIAL TIMEKEEPER
OFFICIAL TIMEKEEPER
Alpina
WATCHES

Alpina
GENEVE

SEBRING
Alpina
WATCHES
OFFICIAL TIMEKEEPER

Honda Racing
HPD
Scott Tucker
Luis Diaz
F. Montagny
P2
95
Alpina
WATCHES
BATTERY TENDER
KENNOL
DUNLOP

Alpina
GENEVE
12 Hours of
SEBRING
12 Hours of
EBRING

TACHYMETER
Alpina
GENEVE
SWISS MADE

Alpina
GENEVE
SWISS MADE

Because of the warm weather in Florida, and the very bumpy track, the 12 Hours of Sebring is seen as an excellent preparation for the 24 Hours of Le Mans. The track may be relatively hard, but with the race not being long, it is the perfect opportunity to test the reliability of the car at the beginning of the racing season.

Alpina and Sebring International Raceway have united for the 57th and 58th annual edition of the legendary 12 Hours of Sebring. As a result, Alpina was named the Official Timekeeper of the event. To celebrate this partnership, three professional Racing watches were introduced. The Alpina 12 Hours of Sebring Automatic Chrono GMT is issued in a highly exclusive limited and numbered edition run of just 158 pieces worldwide. Only 258 pieces Alpina 12 Hours of Sebring Automatic will be produced and lastly, the Alpina 12 Hours of Sebring Chrono, production will stop after assembly of the 558th timepiece.

At the same time, these new 'timekeepers' were the first editions within the new Alpina Racing Collection. The rounded and smooth case shape ensures a perfect fit on the wrist of drivers, while simultaneously meeting their toughest requirements, namely the highest levels of accuracy, readability, shock resistance, sturdiness and reliability. With a diameter of 47mm, these watches feature a black dial with red and silver elements. The raised white and luminous indices ensure visibility, especially important during the night hours of the race. Under the 'hood' of the Automatic Chrono GMT lies a complex and durable automatic movement that offers the ability to keep track of the exact 12 Hours of Sebring, through sophisticated chronograph functionality with a 12-hour counter in the lower half of the dial. To highlight this feature, a subtle red '12' symbolizes the finish line. Furthermore, the watch shows the time and date, and offers a GMT function. Both the front and reverse sides of the timepiece are fitted with ultra scratch-resistant sapphire crystals.

ALPINA RACING
AL-725B5AR26 Alpina Racing Chrono Automatic (2011)
AL-535B5AR26 Racing Automatic (2011)

ALPINA RACING
Alpina is sponsor of Level 5 Motorsports

Alpina
GENEVE
SWISS MADE

THE 24 HOURS OF LE MANS WITH LEVEL 5 MOTORSPORTS

The 24 Hours of Le Mans, also part of the Triple Crown of Endurance Racing, is the world's oldest endurance car racing event still being held; it takes place in France, on Le Circuit de la Sarthe. It all started 89 years ago, in 1923. The idea was to hold a 24 Hours race every year and crown a winner every three years, based on the total performance. This idea was later abandoned and the race evolved in a number of different ways, but the goal was to change the race as little as possible, and to retain its originality, which gave this race its great historical reputation.

Technically, this is probably one of the hardest races, because of its extremely long racing time and its quick variations from racing track to public roads. The public roads are not as smooth and well kept as the racing tracks because of the traffic all year round, which can give the drivers a hard time. Teams must consist of three members who usually do two-hour shifts, thereby giving each driver four hours to rest between their shifts. The 24 Hours of Le Mans can in some ways be compared to the Olympic Games; the race has been held for decades and is about passion. Winning this race is all about honor and pride, not just about money and fame, and that is what we like at Alpina Geneve.

Racing Sports watches remain true to the heritage of Alpina, but benefit from cutting edge precision engineering in their conception, production and testing to ensure that each watch that leaves our manufacture is perfect in every way. It is no coincidence that Alpina proudly joined forces with Level 5 Motorsports, who races in the legendary 24 Hours of Le Mans. Last year, Level 5 Motorsports managed to finish in third place and that success gave the team even greater motivation to win the 24 hours of Le Mans.

Level 5 Motorsports' team for the 24 hours of Le Mans 2012 was made up of Scott Tucker, Luis Diaz and Christophe Bouchut, three men with very different backgrounds, ages and nationalities. However, the shared characteristics that holds these three racers together are their love and passion for cars and for racing, their infatigability and their ambition to win.

With five major victories in the past two years, Team Level 5 Motorsports is doing extremely well for a newly esthablished team. With Scott Tucker at the helm, nothing seems to stop them. But how did this all begin?

In 2006, Scott Tucker, at the age of 51, embarked his racing career by setting up Level 5 Motorsports. He started racing in the Ferrari Challenge for the first two years. In 2007, Tucker claimed his first win; he had talent and ambition, in spite of his age. "Since day one, we have set out to mount a first-class effort toward winning a World Championship." This success gave him the courage to expand the team and

ALPINA RACING
Alpina sponsors Level 5 Motorsport Team 2010-2012

ALPINA RACING
Official Timekeeper of 12 Hours of Sebring

Alpina
GENEVE
SWISS MADE

ALPINA RACING
Alpina is partner RLR MSport

ALPINA RACING
Alpina sponsors RLR Team Racing at the Le Mans series 2008-2009

enter the Rolex Sports Car Series. Over the years the team grew and grew, recruiting new, young drivers, combining experience and passion and always keeping its ambitious spirit. "We've surrounded ourselves with the world's best technical partners, proven crew members and championship drivers. With Alpina Geneve we've found another great partner that shares our vision for excellence."

This year, Level 5 Motorsports upgraded to a new car, the HPD ARX-03b, a racing car developed by Honda Performance Development. The car was developed and built in 2012 and is a more modern version of its predecessor, the Acura ARX-01. It had an outstanding debut, grasping 6 victories out of the first 7 races it participated in. The car has a height of 1030 mm and a width of 2000 mm. With a carbon fibre monocoque and a Honda HR28TT 2.8 Litre V6 Turbocharged engine, it has the perfect balance of strength, weight and power

ALPINA RACING
AL-353B5AR36 Alpina Racing Chrono Quartz Big Date (2012)

Alpina Geneve created a watch especially for the partnership with Level 5 Motorsports, the Alpina Racing Automatic. Featuring a 50 mm stainless steel case with black dial, black leather strap with red contrasting stitching, full 12-hour chronograph function, sapphire crystal with anti-reflective coating and an automatic movement with transparent case back to view the movement, it is the perfect combination of beauty, durability and precision.

ROMAIN BRANDELA

Since 2009, Romain Brandela has been our Alpina Racing Ambassador. He is a non-professional driver in the Gentlemen's Trophy. From a young age, Romain Brandela has been passionate about speed and racing cars. He pursued his dream and became a racer. When Romain Brandela came to Alpina and said that he was a fan of our watches, we were more than happy to sponsor him. Over the past years, Romain Brandela has raced in a number of different cars like the BMW Z4, the BMW B6 and the Lamborghini Gallardo.

RLR MSPORT

Another racing team Alpina proudly supports RLR MSport. Team RLR is a British endurance racing team that is based in Bolton. While it has taken on the challenge of different endurance races in Europe, its ultimate goal has always been to win the 24 Hours of Le Mans. The car that the team currently uses is the MG Lola, a beautiful car that can reach tremendous speeds in a matter of seconds.

ATELIERS DEMONACO

MONACO Views of the Principaute of Monaco

2011
F9
PRINCIPAUTE

INTERVIEW WITH

PIM KOESLAG

ATELIERS DEMONACO
dMc-TBRM-SC-GR Grand Tourbillon Répétition Minute Royale (2010)

PIM KOESLAG
Master watchmaker and co-founder of Ateliers deMonaco

FROM THE DEVELOPMENT OF FRÉDÉRIQUE CONSTANT'S OWN CALIBERS TO THE ATELIERS DEMONACO

It must have been extremely interesting for a young person to be intimately involved in developing a new watch movement.

PIM: It was a tremendous opportunity. I was 21 and hadn't yet graduated from the school of watchmaking in Holland.

Did you always want to be a watchmaker?

PIM: No. When I enrolled at the school, my original intention was to become a goldsmith. In our first year there, goldsmiths, silversmiths and engravers all attend the same class together.

Is there a deeper reason for that?

PIM: Yes, because the first year is devoted to crafts in general. Then I discovered that I like watchmaking. I found it much more interesting and more exciting than the goldsmith's art. Watchmaking combines craftsmanship, creativity, design and technology. That's also true in jewelry-making, but to a lesser degree.

How long did you stay in the same class with students from other disciplines?

PIM: An entire week! Afterwards I was sure that it would have to be watchmaking for me. I knew it would give me great pleasure. At first, we primarily focused on larger timepieces, i.e. clocks.

Did you repair a few clocks or did you do something else there?

PIM: Our school was in Amsterdam. My parents' home was on the eastern side of Holland, towards the German border. I would go home to my parents every weekend, partly because my girlfriend lived nearby. I also needed to earn some money now and then, because Amsterdam is a relatively expensive city. Whenever I had the chance, I would repair clocks at the weekends. Those were my first experiences with the subject.

Had you already acquired the necessary skills?

PIM: Clocks are much larger than watches, and that makes the work easier. But it takes practice and experience before you can work with them completely. I wanted to properly service or repair the clocks that were entrusted to me. After all, you don't get paid in the end unless you've done your work right!

How did you acquire the essential skills?

PIM: When my fellow students went to lunch, I would go to our school's library and learn what I needed to know. My weekend job strongly motivated me to learn.

Did you have enough customers?

PIM: That's interesting too. Word gets around pretty quickly, if you do something well and give customers more than they paid for, for example, if you also tell them about the background of their timepiece. Word of mouth soon brings new clients to your doorstep.

How did you meet Peter Stas?

PIM: My teacher introduced us. During our third year at watchmaking school, we took a trip to Geneva. We were ten students altogether, if I remember correctly. We visited Patek Philippe, Rolex, Vacheron Constantin and Frédérique Constant.

QUALITY CONTROL
Master watchmaker Pim Koeslag checking every detail on a Ronde d'Or Grand Tourbillon XP-1

Why was Frédérique Constant among them? It was in a totally different league at the time.
PIM: Yes, that's what we thought. But our teacher called our attention to the fact that Frédérique Constant had been founded by a Dutch family and had quickly enjoyed success. He said it would be interesting for us to hear their story, so all the students in my class visited the company in Chêne-Bourg. I had always been interested in watch design, so I prepared a few things of my own to show to Peter, whom I hadn't previously met.

Did you bring sketches that related specifically to Frédérique Constant?
PIM: Of course. For example, I brought along a few drawings for Heart Beat models. I had never heard any details about this brand before, but I brought a few designs in my sketchbook.

So you arrived in Chêne-Bourg.
PIM: Yes, we met Peter there and he asked our teacher and us students if anyone among us would be able to help his brand create and develop a movement of its own.

Did you have even the faintest notion about such things?
PIM: Not a clue. We didn't learn that sort of thing in our school. It was an academy for watchmakers, not for designers or movement developers. Courteous silence was the only answer Peter's question elicited.

You remained silent too?
PIM: At first, yes. But when it gradually became embarrassing, I said that I would be glad to give it a try, although I admitted that I had no prior experience in the field.

Peter laughed?
PIM: Not at all. He said, "Great. Let's schedule a meeting to talk about the details." We exchanged ideas about several aspects via email to prepare for the upcoming meeting. Two weeks later, I was on a flight to Geneva. Things became much more specific at that meeting.

Did he commission you for a specific task?
PIM: Yes. Peter shared his already very concrete ideas with me. His thoughts weren't specifically about the technology, but they were nonetheless very precise, probably because Frédérique Constant had already used several basic movements from ETA.

Could you tell us some of the details from that briefing?
PIM: Sure. For example, Peter wanted the balance to be readily visible at "6 o'clock." He wanted something beautiful and clearly different from familiar ETA calibers; not simply an opened movement, but a genuinely open construction that would clearly express the Heart Beat philosophy. I found his idea very stimulating.

How long did it take before you had something to show him?
PIM: First, I must repeat that after graduation from watchmaking school, I was a qualified watchmaker, but not a design engineer for watches. My preliminary sketches were for a balance that was repositioned on the front of the movement at "6 o'clock." I envisioned a fine regulating system from KIF. Along with the balance, the hairspring was also very clearly visible. But this was merely my first idea.

What were the next steps?
PIM: During the evenings, I thought about ways to put these ideas into practice. Meanwhile, Peter had also contacted Serge Nicolin, a graduate from the Geneva school of engineering. Serge developed the first CAD design for the new movement. *(1)*

Did you need to change many details?
PIM: We progressed rapidly because Peter was quite clear about what he wanted. The first suggestions I made were very close to what he had envisioned. Serge and I progressed collaboratively with the design on the computer. Of course, we had to be sure that what we saw there would actually function in three-dimensional reality.

You developed the first prototype from that?
PIM: Exactly.

Were they functional?
PIM: The first prototype basically ran quite well. Needless to say, there are always a few stumbling blocks and we had to push them aside to clear the way. We made some improvements and then we built 10 prototypes, which we optimized further. Afterwards we made 100 prototypes. When there was nothing left to criticize, we began the first zero-series movements.

Could you tell us about the several prototype stages?
PIM: One learns a great deal from the first prototypes, but what's going on in the new movement doesn't become clear until you check the second set of prototypes. You gradually become familiar with the movement's details, its advantages and its quirks. After making improvements based on your experience with the first two prototypes, the third prototype should run almost perfectly. Only then was the road clear for the zero series.

Where did you produce the first zero-series movements?
PIM: On the premises, and mostly by me personally.

Did you use sets of gears from ETA?
PIM: In the beginning, yes. But we gradually liberated ourselves from them. Of course, we cannot make everything on our own premises, nor would we want to do so, but we cooperate with qualified suppliers who can also provide us with gears, pinions and staffs. We no longer need ETA's components.

The time leading up a launch is always very exciting.
PIM: It certainly was for us. Everything had to be ready for Baselworld 2004. Things became really hectic. I drove my car from Geneva to Basel on Wednesday evening and I had the first watches with me.

(1)

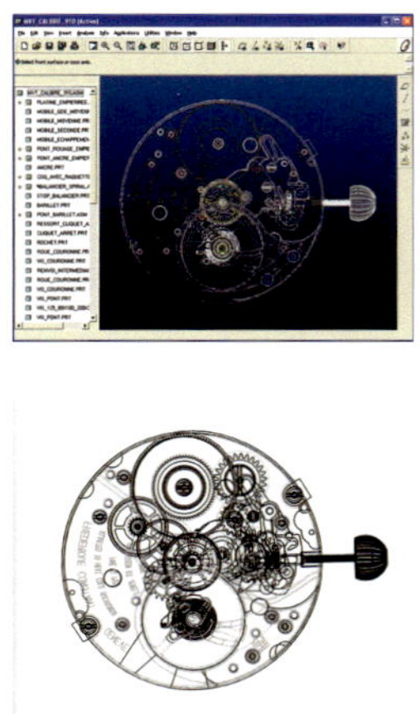

Development of the base caliber on Pro/ENGINEER software

You were already on Frédérique Constant's payroll at this time?

PIM: Naturally. I finished watchmaking school in May 2003. Contractual negotiations with Peter were relatively simple. We're both Dutch. Two weeks later I married my girlfriend and we moved to Geneva.

Do you and Peter speak Dutch with one another?

PIM: Yes, of course. Peter always knows exactly what he wants, but we don't speak exactly the same language: I'm a technician and he's a businessman, yet we understand each other very well. His ideas are always immediately clear to me.

Was this your first extended stay in Geneva?

PIM: No. The student who graduates from our school at the top of his class wins a prize. It's an internship at Patek Philippe in Geneva. All expenses are paid from Geneva. I had that good fortune in January 2003.

I assume Patek Philippe was hoping you'd decide to work for them.

PIM: Exactly. When my internship was over, they asked me if I would like to work for their manufacture, but I told them that I planned to join Frédérique Constant.

How did they respond to that?

PIM: They were incredulous. They said, "Frédérique Constant? Who's that?" My friends and colleagues were likewise perplexed when they heard that I had turned down an offer from Patek Philippe. Almost every watchmaker dreams of working for them. More than one person told me that my decision and my behavior were rather peculiar.

Why did you reject such a prestigious offer?

PIM: I love to create, to design and build things. Working solely as a watchmaker at a workbench isn't enough for me. Watchmaking is more than 500 years old, but many things still remain be discovered, invented and designed. I would have the freedom to do precisely that at Frédérique Constant. Patek Philippe couldn't offer me comparable freedoms and opportunities because I was still a newcomer in the profession. At least 10 years come and go at Patek Philippe before you get a chance to even touch a minute repeater. Peter, on the other hand, recognized my creative potential and invited me to come aboard.

Which brings us back to the FC-910. How were the watches received at the trade fair?

PIM: Our limited series of 500 wristwatches sold out in less than a week.

So you had to produce more?

PIM: The first 500 serially produced watches were likewise primarily my handy work.

Have you ever regretted your decision to come to Frédérique Constant?

PIM: Not for a moment.

The FC-910 was followed by modifications and further development into an automatic movement.

PIM: Anyone who knows Peter also knows that he never lets things come to a standstill. The FC-910 was followed by the FC-915 with moon-phase display, the FC-930 and the FC-935 with lunar indicator. We present at least one new item every year: it isn't always a new movement, but it's always an interesting further development. Production continually grows on this basis. We also do everything we can to continually improve our products. All this must keep pace with the overall development of the family business.

There are several clearly different inputs at Frédérique Constant, from Peter and Aletta, but also from you.

PIM: That's the wonderful thing about Peter. He's always eager to learn about new ideas. He listens carefully and never categorically rejects anything.

How does he handle suggestions?

PIM: His responses differ depending on the suggestion. Sometimes he'll say that we don't have the capacity for a particular idea right now or that it's too risky or too expensive. I had initially thought that it could be feasible, but after thinking it over I usually agree with Peter. He has an outstanding track record. He's right more than 90% of the time! He knows this difficult business like the back of his hand and he has plenty of relevant experience.

In your view, what is Frédérique Constant's secret? Why is this brand able to offer genuine manufacture craftsmanship at such accessible prices, when the same amount of money paid to another brand would buy only a watch encasing a standardized ETA movement?

PIM: We always want to offer our customers something special, but Peter also always keeps the price positioning in the back of his mind. He knows exactly how much money he can ask for something and how much he would like to ask for it. As a businessman, Peter can calculate perfectly. If there's no other option, he'll sometimes even forego a profit margin. We don't give our products away for free, but if you keep a sharp eye on expenses, then "accessible" and "manufacture" needn't be mutually exclusive. Our products prove that they can go together. Astronomical prices don't harmonize with our philosophy.

That surely doesn't explain everything. If it did, many other manufacturers would be charlatans and usurers.

PIM: We save, but in the right places. Take the decoration of the movements, for example. Beauty is a must for us, but only where a viewer's eye can see it, so we don't think it's necessary to put elaborate circular graining on the undersides of bridges and cocks. There's no technical need for embellishment there and no one will notice its absence, but putting it there consumes plenty of cash. One detail is added to another and the final result is a comparatively accessible price.

But you use a relatively costly regulating system from KIF in the FC-910.

PIM: That's what makes this caliber family more costly than the Maxime Automatic, where we use a less costly alternative from Incabloc. Sometimes one asks oneself if KIF is really so much better than Incabloc. Does the difference in quality justify the greater expense? Of course, the client gets a more exclusive product. .

Let's talk about the Ateliers deMonaco, Peter. Watches with this new brand name play in an entirely different craftsmanly league and accordingly sell for higher prices. (1)

PETER: Pim came to me at that time and told me that he had been invited to work for a small brand, which had also offered him shares in the company. I asked if he didn't feel comfortable at Frédérique Constant. He said that wasn't the issue: he felt very happy here, but he wanted to own something.

That was probably a decisive impulse.

PETER: The founding of the Ateliers deMonaco could indeed be a step toward professional independence for Pim. We already had the plans and a few components for a minute-repeater wristwatch, but the complication didn't really make sense for Frédérique Constant because it was far too complex and costly. So I suggested that we found a new firm. Pim could continue working at our headquarters, but he'd also own shares in the new company.

A tempting idea.

PETER: In this way, at age 28, Pim could put his ideas into practice and simultaneously learn how difficult it is to be an entrepreneur.

How did you choose Monaco, of all places, for the headquarters of the Ateliers?

PETER: I was talking with my friend Robert one weekend. He was a marketing director, and he suggested Monaco as a favorable location.

For what reasons?

PETER: Robert told me that there were more than enough watch companies in Geneva. The city really doesn't need another. Monaco, on the other hand, is the world's most luxurious place. Picasso, Monet and many other famous artists were inspired by its special light and unique flair. And in Monaco, we would be close to truly wealthy customers. (2)

(1)

ATELIERS DEMONACO

MONACO
Distinctive and iconic spirit of Monaco is reflected in every dMc timepiece

(2)

INSPIRATION
Paintings of Matisse and Georges Seurat

Can anyone simply go ahead and use "Monaco" as a brand name?

PETER: No, we first had to ask H.S.H. PRINCE ALBERT II of Monaco to allow us to use the name. But that wasn't a problem for a company with its headquarters in the principality.

The air is extremely thin in such a discriminating and pricey category.

PETER: You're absolutely right. Especially in the wake of the financial crisis of 2008, money wasn't spent quite as readily as it was before. That's why deMonaco has remained small and grown slowly. More personnel and larger ateliers wouldn't make sense anyway, because we also need to sell the products.

ATELIERS DEMONACO
Presenting a unique piece to H.S.H. PRINCE ALBERT II of Monaco (2010)

page 292-293
CREATION
Applied index for a deMonaco dial (l), Technical drawing dial (r)

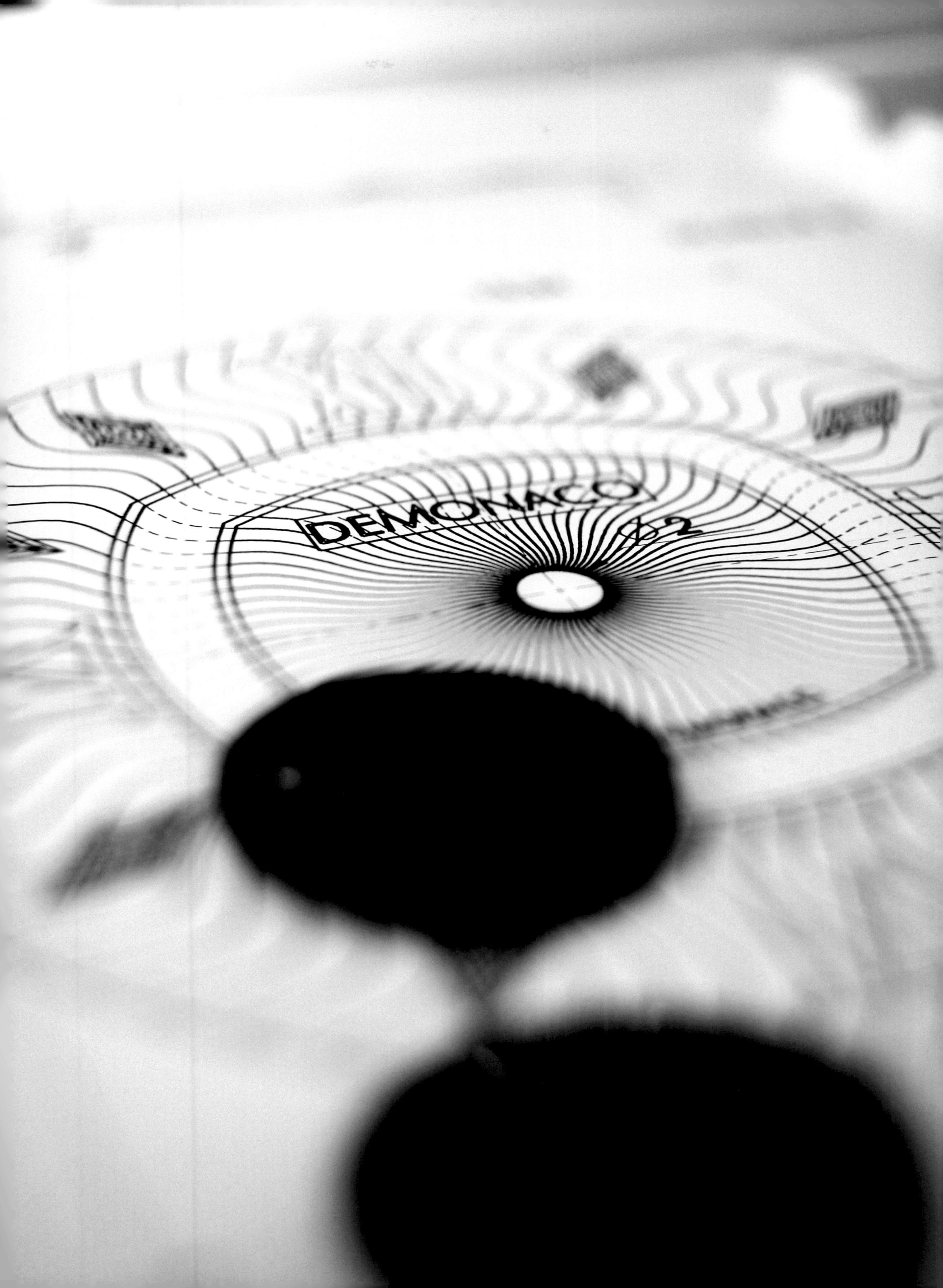
DEMONACO
62

When one looks at wristwatches with the deMonaco insignia, one is immediately impressed by their sensational cost-benefit ratio. Where else, for example, could one find a first-rate tourbillon crafted in such high quality but selling at such a moderate price?
PETER: That's a fact. It's also why we have a few very good customers who regularly buy deMonaco watches. But the production numbers are still quite modest. We make and sell only 60 to 70 watches each year. That's not a very large number. We'll see what the future brings. It's far more important that Pim and Manuel can realize their ideas there. Having their own little world at the Ateliers deMonaco has helped them to understand what it means to be entrepreneurs. Manuel is the CAD/CAM development engineer and co-shareholder of Ateliers deMonaco.

In addition to the tourbillon, deMonaco's spectrum also includes a perpetual calendar and a minute repeater. Which project was most demanding?
PETER: In my view, the perpetual calendar. We were able to master the minute-repeater quite quickly. The production is entrusted to an experienced specialist. He knows this subject down to the very last detail and he can perfectly craft the few specimens with all their necessary fine adjustments and modifications. The perpetual calendar gave us far more severe headaches, but we learned a lot from it.

Also for Frédérique Constant?
PETER: Yes. In his capacity as technical director, Pim can now implement what he learned from those experiences into future developments at Frédérique Constant. This excursion into the realm of complications definitely makes sense. It also earns some revenue so we can at least partially cover our costs.

Rents in Monaco are exorbitantly high.
PETER: Our atelier in Monaco is quite small. We can afford it. But unpleasant things sometimes happen there which wouldn't have occurred if I had monitored the situation more closely. For example, an Italian concessionaire purchased four watches which he then immediately passed along to a customer in the Middle East. The customer promptly paid the sizeable sum to the concessionaire, but thus far we've received only a small part of it.

PIM: I'm very proud of the Ateliers deMonaco.

Congratulations, Pim, you're a co-owner of the Ateliers deMonaco. How do you view the undertaking?
PIM: When we launched the Frédérique Constant Tourbillon on the market, I was already working on the next project, which I had chosen because I wanted to give myself a very special challenge.

HEAD QUARTERS
Principal Atelier & Office Ateliers deMonaco in Palais De la Scala in center of Monaco

What did it involve?

PIM: It was a movement with both minute-repeater and tourbillon complications. Strictly speaking, I had initially only planned a classical minute repeater and then I wanted to augment it with the new tourbillon. So I developed it, mostly in my free time. That led to a prototype, which I showed to Peter.

What was his response?

PIM: Peter was very pleased with it. Not only did it work perfectly, it also resounded with marvelously beautiful chimes. But Peter felt that it wouldn't be appropriate for Frédérique Constant because the price of such a complicated wristwatch would necessarily far exceed the limits of the brand's accustomed price segment. That's a totally different league. The gap between a tourbillon for about 30,000 euro and this complication, which we would have to sell for more than 100,000 euro, could never be filled with credible models.

But that didn't put an end to the idea.

PIM: No. Peter discussed it over the weekend with his Dutch friend Robert, who worked in the marketing department of a big international corporation in Geneva.

The results were obviously positive.

PIM: Yes, indeed. Peter told me on Monday that we definitely must do something with this development, optimally under a new brand name with its headquarters in Monaco. He also had a clear concept about who would play which roles. Robert would take care of the marketing aspects, the onsite premises and everything else that would be needed to establish the company, while Peter and I saw our tasks as identifying in the products that would be necessary. Peter also suggested ideas about what would be feasible in the area of haute horlogerie to round out the upper end of Frédérique Constant's spectrum.

Did you both share the same opinions?

PIM: Yes, I believe so. My suggestion was a perpetual calendar as a logical addition to our palette of complications, which thus far contained a tourbillon and the new minute repeater.

How did things develop after that?

PIM: The new Ateliers deMonaco brand and its significantly upgraded tourbillon debuted in 2009. This was followed relatively quickly by our second complication, i.e. the minute repeater with tourbillon.

ATELIERS DEMONACO
Inspecting a Ronde de Monte-Carlo

Many watch brands offer tourbillons. What's special about yours?
PIM: To explain that, we first need to talk about the tourbillon that Frédérique Constant launched in 2008.

A tourbillon with many special features...
PIM: Including a system to perfectly poise the rotating cage. *(1)*

How did this project come about?
PIM: One day Peter came to me and said that he wanted Frédérique Constant to have its own tourbillon. When I heard that I had to admit that I didn't yet have the relevant experience in designing and engineering a tourbillon. I had worked on our Progress STT calibers, but that was a long way from a construction of our own. My work on those calibers, however, had enabled me to gain some experience which shouldn't be underestimated: I knew exactly how not to design and engineer a tourbillon! I used that as a basis for figuring out how it could be done much better.

Could you give us an example?
PIM: Our tourbillon's rotating cage was supported on only one end. I believe that a so-called "flying" tourbillon always involves a risk. Certain problems are involved with the use of ball bearings. That's one aspect. Another has to do with the tourbillon's power source. It seemed to me that if we used a different gear-train we could make the tourbillon rotate more uniformly and more efficiently. The third factor was a perfectly poised cage.

Combining all these important aspects significantly improved the traditional tourbillon.
PIM: Yes and it also brought us several big steps forward in our deMonaco project. We produced about 100 movements for Frédérique Constant, but in my leisure time, I strove to further improve the caliber's rate performance. The FC-980 needn't fear comparison with our competitors' products, but I knew that further improvements could still be made. The cage is very well poised and we use Nivarox as the material for the balance and its hairspring, so there's no reason why our tourbillon should run less accurately than a movement that has earned an official chronometer certificate.

What exactly did you achieve?
PIM: The daily rate values of the deMonaco version were between minus one second and plus two seconds, That means they're significantly more accurate than those specified by the official chronometer standard.

(1)

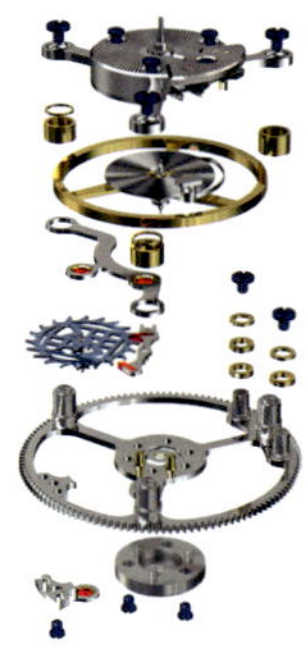

CALIBER
Tourbillon Cage exploded view

SILICON
Escapement wheel in Silicon

(1)

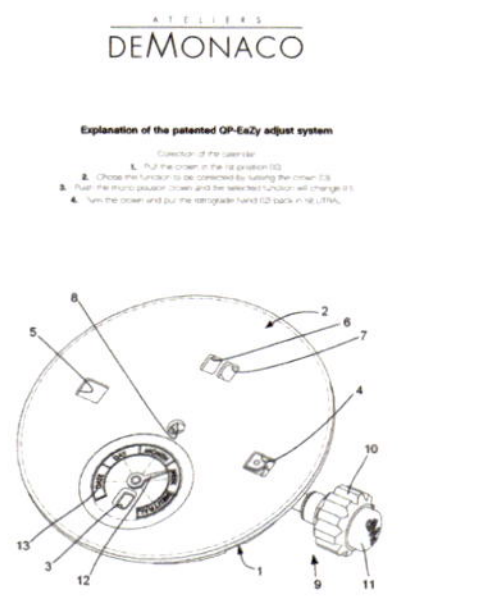

PERPETUAL CALENDAR
Drawing from the Patent
of the QP-EZ Caliber

What was your trick?

PIM: There was no trick! Nothing but concentrated and painstaking work, step by meticulous step. I learned a great deal in the process. To finely adjust the tourbillon per se, I developed a movement in which the entire mechanism runs, but the cage doesn't turn. This enabled us to easily adjust the balance and its hairspring, just as one can do in an ordinary movement without a tourbillon. That's perfect! As far as I know, there had never before been such a tool. There's really no end to the learning process.

Was the innovative perpetual calendar also originally intended for Frédérique Constant?

PIM: Yes, because this complication had been lacking in our collection. It could have augmented our tourbillon, which likewise occupied the high end of Frédérique Constant's collection. On the other hand, we had set ourselves the task of always creating something new and off the beaten path for the Ateliers deMonaco.

Entirely new varieties of complications?

PIM: No, not that. What we have in mind are existing additional functions, e.g. a perpetual calendar, which we then build in a different and better way. As I explained, that's what we had already done with our tourbillon. The minute-repeater likewise had certain unique features. The same would have to apply to a perpetual calendar for the deMonaco collection. (1)

Could you cite two examples?

PIM: Sure. First, we have easy, totally uncomplicated setting and adjustments which we call "EaZy Adjust." Such a helpful feature had never before existed in this technical realization. Then we have the calendar's displays, which jump ahead within a very brief time span each night at midnight. Innovations such as these reflect the spirit of the Ateliers deMonaco. Whenever possible, we take something that already exists and tangibly improve it.

ATELIERS DEMONACO
dMc-TB-RR-GR Ronde d'Or
Grand Tourbillon X^{2}1
in 18K Rose Gold (2009)

A minute-repeater ranks amongst the most intricate mechanisms in fine watchmaking. Scarcely any other complication is more complex.

PIM: That's right. The interactions among the numerous components are extremely complex. I collaborated with highly experienced watchmakers, but developing the minute-repeater nonetheless required plenty of time. Regulating the sequence and the acoustic aspects was an especially tremendous challenge. I frequently needed to discuss the details with experts. You can't simply wrap the gongs around the movement and expect a pleasantly pitched chime to ring. Special techniques are involved and they demand ample experience. Nothing happens by serendipity here.

Had you already acquired relevant know-how at the school of watchmaking or during your internship at Patek Philippe?

PIM: Not at all. I had seen and heard a few minute-repeater watches, but that was the extent of my familiarity with them. During the developmental phase, I visited every watch shop in Geneva that carried minute- repeaters. I strapped them to my wrist and listened to their striking mechanisms. That's how I learned to separate the wheat from the chaff. There are also a few internet sites where one can experience these watches, including how to regulate the interval between successive chimes. In a nutshell. I had to conduct plenty of research, before I could begin to develop our own minute-repeater. There was no other alternative. I read many books. I talked with watchmaker colleagues. And I gradually formed an opinion about what might be appropriate for the Ateliers deMonaco.

Comparatively soon after completing your training as a watchmaker, you dared an ascent to watchmaking's highest summit, i.e. one of the most complex of all complications. Were you ever afraid that this project might be beyond your abilities?

PIM: No, never. Incidentally compared to our perpetual calendar, the minute-repeater isn't really very much more complicated. I would estimate that the calendar is about 90% as complex as the minute-repeater. Some components of the calendar mechanism are even more complicated than the striking mechanism.

Why is that?

PIM: Our movement with perpetual calendar consists of approximately 350 parts. More components are needed for a minute-repeater. We're talking about nearly 400 parts here. In retrospect, I can see that it wasn't simple, but I never had the feeling that I wouldn't be able to handle it.

Is there a fundamental difference between developing something entirely new and elaborating on an existing mechanism?

PIM: For me, it's much better and more interesting to begin with the preliminary sketches. That's why one can identify oneself with every step. One knows every detail and has a clear idea of where one wants to go. I wouldn't want to forego the learning processes. There were so many nights when I couldn't fall asleep because my mind was preoccupied with the construction. It's truly adventurous to create an entirely new mechanism from scratch.

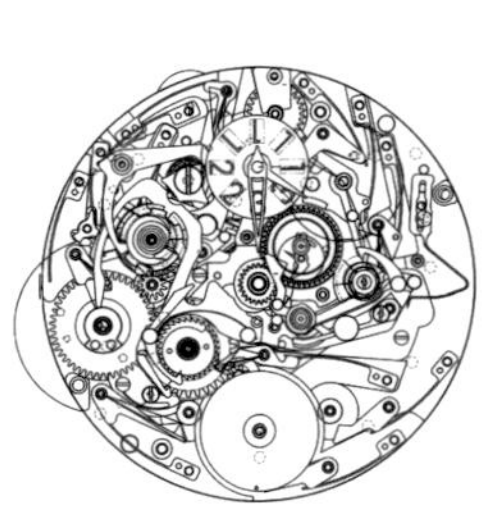

PERPETUAL CALENDAR
Technical drawing of QP-EZ
Caliber under dial view

ATELIERS DEMONACO
Caseback of dMc-TBRM-SR-GR
Ronde d'Or Grand Tourbillon
Minute Repeater

Did you develop special gongs for the minute-repeater?
PIM: Only three companies in the watch industry make gongs, and they supply 90% of all manufacturers of repeater striking mechanisms. We collaborated with one of them for the production of our gongs, so I really can't describe it as an entirely new invention. We followed a classical path toward the best possible acoustic experience, but the blend of innovation and tradition is totally in harmony with our manufacture philosophy at the Ateliers deMonaco. For example, on the one hand, we integrate modern aspects with an innovative silicium escapement and bridges made from sapphire crystal; on the other hand, we use traditional black polishing on the steel parts and time-honored manual engraving on the gold rotor.

You said that a perpetual calendar is nearly as complex as a minute-repeater. Could you explain that?
PIM: The 60 to 70 components of a perpetual calendar are usually mounted on their own plate, which is then affixed to the front of the existing movement. The displays are set and adjusted by pressing buttons on the flank of the case. Practically no changes are needed in the basic movement. The daily switching sequences occur over the course of several hours.

But you didn't want that.
PIM: Exactly. Before we started the developmental process, we again asked ourselves precisely what we could do differently and better. We identified several details in the calendar that we wanted to improve. First, we wanted to enhance the legibility of the displays by enlarging the apertures in the dial. Then we wanted the five calendar indicators, including an unconventional display for the calendar week, to advance speedily at midnight. Furthermore, all of our calendar displays can be set via the crown. We have a patent on that. Last but not least, you can reset the hands either forward or backward at any desired time of the day or night without damaging the mechanism. At that point, it should have been clear that we needed to include the basic movement in the construction. That of course, involved significantly more labor than usual. (1)

(1)

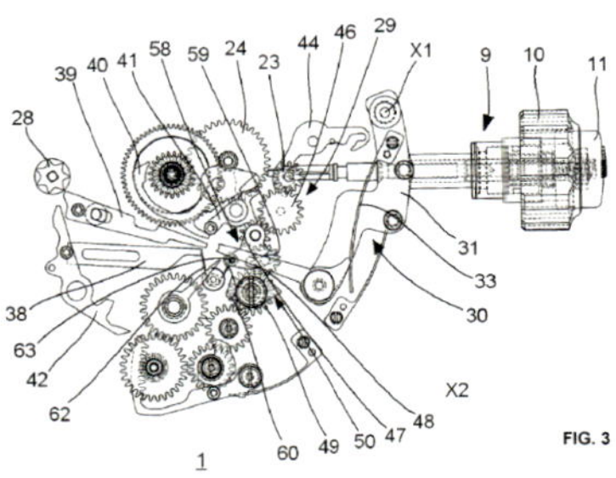

FIG. 3

PERPETUAL CALENDAR
Drawing from the Patent
of the QP-EZ Caliber

It's now clear to me that this construction and the costs of manufacturing involved were too high for Frédérique Constant. Could you imagine reducing the complexity, for example, by simplifying the perpetual calendar into an annual calendar?
PIM: We could do that. Peter and I love challenges, and they could involve commercializing the calendar. But right now, we want the version from the Ateliers deMonaco to demonstrate that we can indeed accomplish state-of-the-art horological feats. Needless to say, we've also thought about subsequent and perhaps simpler variants. They would feasible without causing any problems and could be achieved in a relatively short span of time.

Were these your ideas or did Peter contribute his?
PIM: Peter always thinks and acts cleverly. He looks ahead. His credo is to learn from complexity and then reduce the degree of difficulty. That's more advantageous than trying to advance from the simple to the complex.

That sounds quite plausible.
PIM: There's something else too that we shouldn't forget. If we had launched the perpetual calendar with Frédérique Constant's insignia and if we had made, let's say, a series of 1000 such watches, we would have faced a mountain of problems. The mechanism is so elaborate that we could never have produced those quantities within a reasonable timeframe. It would have made absolutely no sense at all.

It would have been possible with more personnel.
PIM: It might have been possible to hire additional watchmakers, but each of them would first need to master this highly demanding specialty. A training period of at least one year would have been needed before they could have begun working on production per se. No, we didn't even want think about that. (1)

(1)

Rotors of Ateliers deMonaco are made by hand in 18K gold

There's good reason to eagerly look forward to what the future will bring. How did potential customers respond? After all, deMonaco isn't yet very well known.
PIM: Their response was surprisingly positive. But a perpetual calendar needs a lot of explanation before customers can fully comprehend what makes this wristwatch so special and just how much horological creativity and complexity are built into it.

Are those the only complications that we can expect from deMonaco for the time being or are other developments in the pipeline?
PIM: We have a long list of ideas and we're presently evaluating their feasibility.

En l'Honneur de S.A.S. Prince Albert II de Monaco
dMc Grand Tourbillon XP1
1/1
ATELIERS DE MONACO
OR - 750
Titanium Gr.5
Pièce Unique

ATELIERS DEMONACO
Unique piece manufactured for and offered to H.S.H PRINCE ALBERT II of Monaco

What about a chronograph? Is a "time-writer" already beginning to appear on the horizon at Frédérique Constant?

PIM: In this instance, we indeed reversed direction. We didn't start with something complex and then simplify it. Our chronograph was specially conceived for Frédérique Constant. To reduce costs and to simplify servicing, we want it to be as simple as possible, yet nonetheless fully functional. The Ateliers deMonaco will surely continue to work on this development. But it's still too early to speculate about what direction this project might ultimately take. We're talking here about the more distant future, perhaps the year 2015. You mustn't forget that we've only been in existence since 2009. Rome wasn't built in a day!

The Ateliers deMonaco's collection also includes a simple self-winding wristwatch.

PIM: Yes, we introduced it in Basel in 2011. Its special features include a manually engraved rotor made of solid gold. The decorations also embody the utmost in craftsmanship. This watch's buyers can admire manual guilloche whenever they glance at its dial.

RONDE D'OR
dMc-MC-LG-WGD (left)
& dMc-MC-LG-RGD (right)

MONACO

Monaco Crest and Flags

DEMONACO DIAL
Inspection of a Ronde de Monte-Carlo dial

Who buys such luxurious wristwatches?
PIM: The automatic model is equally popular with men and women. The cases aren't overly large and the look is characterized by tasteful elegance. Ladies are particularly fond of the versions with diamonds.

You are financially involved in the Ateliers deMonaco. How does it feel to be a young entrepreneur?
PIM: It was and still is a dream come true for me, but I admit that it isn't always easy. After all, I also work for Frédérique Constant and there's always plenty to do there.

How do you divide your time between Frédérique Constant and deMonaco?
PIM: About half and half, I suppose. I'm technical director at Frédérique Constant and Alpina. Combining the two isn't the easiest thing in the world. After all, we annually produce about 120,000 watches. The annual total was just 30,000 watches when I joined Frédérique Constant nine years ago. The whole office then was no larger than our conference room today. Very much has happened in the interim. My team now numbers 15 people. We'll produce about 10,000 manufacture movements in 2012. At present, our upward trajectory looks unstoppably strong. I'm very proud to be part of this success story.

Are you satisfied with the situation, even though you have to stand, as it were, on two separate islands, or would you prefer to work solely and completely for the Ateliers deMonaco?
PIM: I'm very satisfied with my current twofold situation because I love both task fields. Every day, Frédérique Constant makes me acutely aware of the tremendous importance of uncompromisingly high quality. Our philosophy permits no fundamental difference between the few watches bearing the name "deMonaco" and the many thousands of watches with the names "Frédérique Constant" and "Alpina" on their dials. In the final analysis, all our customers must be satisfied with their wristwatches, no matter how much the timepiece may have cost. It's a bit like doing the splits. It isn't easy, but it raises our awareness of problems that we need to avoid in the future. My conception of what it means to be an independent entrepreneur has also totally changed. Very few people have the good fortune to be able to experience both sides as I'm currently doing.

How do you proceed with new developments, regardless of whether they're for Frédérique Constant or the Ateliers deMonaco?
PIM: Basically, there haven't been many changes over the past several years. Peter, Aletta or someone else suggests an idea. We check it, discuss it and finally decide for or against it. If our decision is positive, then the R&D department gets involved. They form a concrete plan, construct, make prototypes, etc.

How often are you in Monaco?
PIM: Quite regularly. That's indispensable. And the journey between Geneva and Monaco is surely not the worst thing a traveler could experience! It takes only 45 minutes to fly from Geneva to Nice, where I rent a car.

Do you invite existing and potential concessionaries to Monaco?
PIM: Naturally, because we want them to experience the flair of the workshop. Afterwards, they can convey our philosophy of craftsmanship and our product philosophy much more effectively to retail customers. (1)

How many concessionaries do you have around the globe?
PIM: About 10, I believe. It's a very small and exclusive circle. We try to get nothing but the best, so we choose with great deliberation and foresight. Slow growth is our underlying idea. You only get one chance to begin. If it goes awry the first time, things can become very difficult afterwards.

Where, for example, are you represented?
PIM: At the Hôtel de Paris in Monaco and at two points of sale in Porto Cervo in Sardinia. We also have one store in New Delhi, one in Bangkok, one in Russia, one in Geneva, etc. They're small but always very fine, just like the Ateliers deMonaco.

Are you personally acquainted with H.S.H. PRINCE ALBERT II of Monaco and his consort?
PIM: Yes and I'm very proud of that. I was invited to the palace in 2011 and I had the honor of personally delivering his deMonaco wristwatch – a tourbillon.

How did he happen to choose that one?
PIM: We sent a catalogue to H.S.H. PRINCE ALBERT II of Monaco and we invited him to choose based on his personal tastes and interests. That's exactly what he did. We made him a one-of-a-kind timepiece in accordance with his personal preferences.

DEMONACO
MADE IN MONACO

LA SIRÈNE
Unique Piece inspired by Mrs Charlène Wittstock, Her Serene Highness The Princess of Monaco

H.S.H. PRINCE ALBERT II of Monaco visiting Ateliers deMonaco at Top Marques 2010. Her Serene Highness The Princess of Monaco inspecting one of the watches.

What about the princess?
PIM: Princess Charlène got a special ladies' automatic with blue gems from South Africa. It was the first feminine model from the Ateliers deMonaco. (1)

Could you describe your meeting with the prince and princess?
PIM: It was a very important and sublime moment for me. We spent about half an hour with them. Albert was interested in our watches and our production. That was a delightful and unexpected surprise.

Did you meet H.S.H. PRINCE ALBERT II of Monaco afterwards?
PIM: The prince has an amazingly good memory. Each time we meet, for example, at an exhibition, a trade fair or some other event, he shakes my hand and asks about the progress of our business. I would never have expected that.

(1)

How is the atmosphere in the principality?
PIM: It's very personal, almost like a family. People know us in Monaco. Our work is appreciated there. And the fact that the prince and princess own watches from our brand naturally has a positive effect on our business development.

Are there other deMonaco watches with a close connection to the principality?
PIM: We made a few watches to commemorate the 1956 Grand Prix – one-of-a-kind timepieces in response to individual orders.

Gérald Genta lived in Monaco. Did you occasionally meet him?
PIM: Yes, and that too makes me proud. He visited us at various exhibitions and he took a long, thorough look at our collection. At first, however, he was a bit skeptical that we had chosen Monaco, of all places, for our headquarters. *(1)*

Did Gérald Genta also visit your workshop?

PIM: Yes, I had invited him. He came and stayed nearly four hours. We talked about his famous watch designs. Another time we were both at a dinner in the Hôtel de Paris. When I add up all the time we spent together, it comes to almost an entire day. Toward the end, we began thinking about a mutual project for the future, but unfortunately Gérald Genta died two weeks later.

Did he make any comments about your watches?
PIM: That's a third source of pride for me. Genta liked our watches. He praised their harmonious design and the meticulousness of the craftsmanship.

Who designs deMonaco watches?
PIM: Madeleine from our design department and I. These watches are really my first designs. That's all the more reason why praise from Gérald Genta means so much to me.

You're still a very young watchmaker and businessman, but you're already enjoying strong success in both métiers. Is this where you see your future or do you envision yourself at one of the big traditional manufactures such as Patek Philippe or Vacheron Constantin?
PIM: One never knows what might happen, but I appreciate the fact that within the limits of feasibility, I can make my own decisions, accept responsibilities and face big technical challenges. In this regard, I can optimally evolve and grow here. There's no reason for me to go elsewhere or to look longingly at other manufactures.

(1)

You graduated from a school of watchmaking. Did you learn enough there to cope even moderately well with the tremendous spectrum of tasks that you face here?
PIM: Not at all. You learn the basics there. As far as the rest is concerned, you either have to teach yourself or learn it from life and firsthand experiences. Peter is a shining example of this for me. Whenever he turns his attention to something, in the background is always his hope that he'll be able to learn from it. Anyone who isn't willing to learn will never evolve and progress.

How many of your basic ideas do you borrow from books?
PIM: I own plenty of books and I leaf through them time and again. I always discover something new and

(1)

Reumpel Clock (17th Century)

perhaps usable. It's especially worthwhile to look at the pictures. I don't read quite so much because I find that somewhat difficult. I was dyslexic in school and watchmaking school, with all of the attendant negative symptoms.

Are there watchmakers in your family tree?.

PIM: No, not really. At least, that's what I had always believed. But then something peculiar happened. I was in the library at watchmaking school when I stumbled upon one of my ancestors' family names in a book. Until that moment, I would never have believed that the Koeslags had ever had anything to do with horology. After further research, I discovered that my great-great-great-grandfather was a clockmaker. His family name was Reumpel and he was a famous Dutch clockmaker in the 17th century. His clocks looked quite ugly to me, but their quality was high and they were therefore relatively expensive.

How did you discover the connection between Reumpel and Koeslag?

PIM: The first clues came from our family archive. Then we continued our research and traced the exact familial relationships.

Would you like to own a Reumpel clock?

PIM: With regard to our family's tradition, yes, I would. But they're far too expensive for me. (1)

If you become and remain a successful watch entrepreneur, then you ought to be able to afford one someday.

PIM: I hope so.

To return to your family: how is your wife?

PIM: After nine years of marriage and two children together, we've established ourselves here in Geneva. It isn't always easy to balance family life and a multifaceted career, but we succeed pretty well.

Is it very different to live here rather than in the Netherlands?

PIM: Definitely. There's somewhat more sun and somewhat less rain here than in Holland. We like that. We also enjoy the somewhat higher quality of life here. An important part of life in France, where we bought a house, is to know how to lean back and enjoy oneself. Things tend to be much more ascetic in Holland.

With so much work, do you also find time for hobbies?

PIM: I would like to have hobbies, but I don't have the time. That's why my two biggest hobbies are my family and my work. I commute between the two in my car, which is a great source of pleasure for me.

INTERVIEW WITH MANUEL DA SILVA MATOS

MANUEL DA SILVA MATOS
R&D Manager and Share Holder
Ateliers deMonaco

Manuel, I'd like to talk with you about the processes that occur at Frédérique Constant and Ateliers deMonaco when you, as R&D Manager, collaborate with Peter Stas and his team to develop a new movement. But before we do that, could I ask you about your previous career?

MANUEL: After completing my education, I began my career at Bunter S. A. in Versoix. Bunter is quite versatile. It's an all-rounder with particularly strong competence in the fabrication of cases lavishly decorated with gemstones. The next phase of my career was at Franck Muller Watchland, where I was responsible for developing the cases for the "Conquistador" and "Long Island" ranges. That was an exciting time! Anyone who knows the convex-concave case of the "Conquistador" will understand why. Pim Koeslag hired me at Baselworld in 2008 and I began working for Frédérique Constant.

Now you not only design and engineer cases, but also entire movements. And you work in a totally different price segment than you were used to at Franck Muller.

MANUEL: We cannot afford to make errors at Frédérique Constant. Wholly different laws apply in this price segment. Compared to other manufacturers, we have a lower ceiling, so to speak. That has its disadvantages, but it's also quite challenging. We must work with the utmost efficiency and exactness because our work will ultimately pay our salaries.

What was your first project for Peter Stas?

MANUEL: Caliber FC 335.

Is that the automatic caliber with a date display and moon-phase indicator which ticks for example, in the "Maxime Heart Beat Moon Date"?

MANUEL: Yes, exactly. Peter assigned us the task, so Pim and I rolled up our sleeves. We were joined later by Hugo, who accompanied us and contributed to the project until its completion.

You, Pim and Hugo have remained a close-knit team ever since.

MANUEL: Pim is a talented watchmaker, Hugo is a perfectly organized production manager, and I love to tinker, to design and engineer things. But we're only three of the many people who collaborate to create a new caliber. Above all, there's our mastermind Peter, who always has the last word. Peter often collaborates on the design and also contributes ideas.

What are you working on right now?

MANUEL: Peter suggested that we develop a "Worldtimer" caliber for Alpina, so Pim and I began thinking about it. As always, I scribbled copious notes. My notepad is like my second memory. It's where I sketch and design things and jot down ideas and approaches to potential solutions. During the next steps, we meet in progressively larger groups. Hugo, Peter and Pim are involved, of course, but we're joined by our head of communications Yasmina Pedrini and several others. Thorough brainstorming is the purpose of each meeting.

Continually sharing information about the most important progress and problems is essential in this phase, isn't it?

MANUEL: Yes, we really can't talk with one another enough during the developmental phase. A golden rule of design engineering states that you should intensively use the conceptual phase, when iterations can be made in far less time and at a much lower cost. Planning is repeatedly modified on the basis of the quality tests. Hugo, Pim and I constantly make modifications based on quality analyses conducted by ourselves and our suppliers.

You said that Peter always has the last word with regards to the final okay. Was that also the case with the Alpina "Worldtimer" caliber?

MANUEL: The new caliber was designated as "AL-718." We provided four suggestions for its perfect integration into the existing caliber structure, and Peter ok'd one of them.

The pace accelerated drastically afterwards?

MANUEL: That's the phase when the design engineer, the participating watchmaker and the production manager must communicate intensively with their in-house colleagues in prototype construction and with external suppliers. Every tiny detail can be relevant, and sometimes the improvements involve mere hundredths or even thousandths of a millimeter. We've developed very good relationships with our suppliers over the years, so they're involved in the developmental process right from the start because they actively and enthusiastically help to improve our products.

DEVELOPMENT
Discussion about the QP-EZ Caliber's plan

Are those the most stressful phases in the developmental work?

MANUEL: The biggest stress comes during the prototype phase. That's always been so and I expect it to remain so in the future. One simply cannot plan and design things in advance with absolute perfection. Even in today's high-tech era, a gap always separates theory and practice. The brutal phase usually lasts about 10 days. After that, the worst is over and things quiet down somewhat. When the supplier has likewise ok'd the manufacturing process, Peter symbolically presses the "start" button to begin serial production.

How long did the theoretical designing and engineering of the deMonaco "Tourbillon" caliber take?

MANUEL: We had the first prototypes after a bit more than three months. We could build on a pre-existing caliber, so the work went relatively quickly compared to a completely new development. Things don't move so speedily when I need to develop something entirely new. Right now, I'm simultaneously also developing our chronograph caliber and an interesting calendar mechanism. Plenty of water will have flowed down the Rhône before the first prototypes are finished!

What do you do to shake off the stress? How do you relax and refresh your energies?

MANUEL: My big hobby is golf. It perfectly balances my professional work. I'm incredibly fascinated by the constantly repeated challenge of propelling a small ball across a large distance and finally into an equally small hole.

May I ask you your handicap?

MANUEL: After twenty years of playing, I currently have a handicap of nine.

Wow! Thank you for your valuable time and for this interesting conversation.

DEMONACO
MADE IN MONACO

GRAND TOURBILLON

WITH MINUTE-REPEATER
THE AUDIBLE DIMENSION OF TIME

ATELIERS DEMONACO
dMc-TBRM-SR-GR Grand Tourbillon Répétition Minute Royale (2010)

DEMONACO

HEARING RATHER THAN SEEING

"Time is what we read from a clock." Einstein used this phrase to describe mankind's most precious possession. His definition resulted from insights articulated in his theory of relativity, which he proposed in 1905. In a thoughtprovoking experiment, the ingenious physicist further explained the relativity of our notion of time: "Time passes quickly if you spend two hours sitting beside a pretty girl, but slowly if you spend two minutes sitting atop a hot stove." A man would seldom or never glance at his watch in the first situation, but would probably consult its dial frequently in the second instance. Looking at a watch is only one way to determine the current time. Those of us who can afford the luxury, can also hear rather than read the time. Haute horlogerie as practiced by Frédérique Constant in its Ateliers deMonaco adds an audible dimension to time that transcends the quiet ticking produced by all mechanical timepieces.

This is easier said and written than done. The minute-repetition we're talking about here ranks among the most complicated additional functions that can be added to a mechanical timepiece. If you want to hear the current time, you need only shift the position of a little slide in the left-hand flank of your watch's case. This triggers a series of chimes. First comes, one or more chimes corresponding to the current hour; next, one double chime for each full quarter hour that has elapsed since the last full hour and third and finally, one chime for each minute following the last full quarter hour. Of course, the user must count each individual chime. If, for example, he triggers his repeater watch at 12:59, it will obediently produce 32 tones, including twelve low notes; then three double chimes, each consisting of a low note and a high one; and finally fourteen high notes. Approximately 100 components must perfectly interact to assure that all this functions properly, i.e. that no chime is left out nor any chime added to those which are absolutely required. If any component fails to play its role, the watch remains silent.

ATELIERS DEMONACO
dMc-TBRM-SC-GR Grand Tourbillon Répétition Minute Royale (2010)

BEAUTIFULLY COMPLEX

The watchmakers in the Ateliers deMonaco are familiar with every detail in the small and very fine hand-wound caliber TB-RM 1 with 48 hours power reserve and in the XP 1-Tourbillon with an escapement made from silicium components.

The greatest challenge is the striking mechanism, which must ring chimes that are pleasantly loud and mellifluously tuned. The first precondition for precisely repeating the time is the synchronization of this mechanism with the hands. To achieve this synchronicity, installing the hands is the last task performed in the Ateliers deMonaco before the movement is encased. The watchmaker triggers the repeater mechanism, counts its chimes and finally positions the hands accordingly. Specially shaped and calculated cams convey the temporal information from the movement to the striking mechanism. An hour snail is responsible for striking the full hours. This helical construction with 12 steps is installed where the hour-hand will later be affixed. Together with the hour hand, the hour snail completes two rotations per day. The repeater mechanism doesn't distinguish between daytime and nighttime hours, but this is not particularly relevant.

DEMONACO

When the striking mechanism is triggered, a little pin touches the appropriate step on the snail. As the number of hours increases, so too does the height of the step. For example it's the eighth step's turn between eight o'clock and 8:59. Now the so-called "rack hook" comes into play. It moves into the eighth gap in its gear-rack, where the striking mechanism's power source propels this rack as far as it will go, and a little hammer delivers eight successive strokes against a low-pitched gong wrapped around the movement. A steel spring lifts the hammer off the gong immediately after each stroke, thus enabling the gong to vibrate freely and resonantly.

The quarter-hour snail turns with the minute-hand. This snail has three double cams -- one each for the first, second and third quarter of the hour. (A fourth cam is unnecessary because the minutes ring immediately after each full hour.) Mechanical scanning occurs after the final hour bell has struck. Then the striking mechanism briefly lifts two hammers in rapid succession. The first mallet strikes the high-pitched gong: then its counterpart hits the low-pitched gong. The minute snail is shaped like a star with four arcing arms, each bearing fourteen tiny teeth. It rotates with the minute-hand and begins to operate only at the end of each chiming sequence. Its teeth represent the maximum possible 14 minutes after the most recently elapsed full hour or quarter hour. Each tooth triggers the hammer to strike once against the high-pitched gong.

Needless to say, if successive chimes ring too quickly one after the other, the user will find it difficult to correctly tally them. A silently operating centrifugal regulator limits the number of strokes to about one and a half blows per second. An interval of 20 to 25 seconds is thus required to complete the series of chimes that corresponds to 12:59 a.m. or p.m.

ATELIERS DEMONACO
Close-up dMc-TBRM-SC-GR
Grand Tourbillon Répétition
Minute Royale

TITANIUM AND GOLD FOR HIGH-QUALITY SOUND

The high quality of the sound produced by the "Grand Tourbillon Minute Repeater" from the Ateliers deMonaco is partially achieved by perfectly matched hammers and gongs, but the case's construction is important too. The high density of solid precious metal tends to muffle chimes. A composite case of titanium and gold solves this old problem. It makes no difference whether the case is rectangular or round. Whatever its shape, its bimetallic composition allows the audible dimension of this unique timepiece to resound with unparalleled richness.

This choice wristwatch is a joy for the ears and the eyes. Viewed from the front, its tourbillon whirls beneath a sapphire crystal bridge and its skeletonized dial offers a clear view of the cadrature for the repeater mechanism. If you want to observe the hammers striking against the gongs, you'll have to slip the watch off your wrist. The back view reveals the full beauty of a gorgeously finished and decorated movement.

A COMPLICATION WITH A LONG HISTORY

Timepieces with repeater mechanisms essentially became obsolete with the dissemination of electric light. The original purpose of this elaborate, costly and correspondingly rare complication was to announce the time during night hours. Such timepieces have always been quite expensive.

Their story began in England in 1687, when a battle raged over a patent to which two parties claimed the rights. Patent protection for a movement that repeated the quarter hours was sought by watchmaker Daniel Quare (1649 - 1724). Its other claimants were the parish priest Edward Barlow (1636 - 1716) and the watchmaker Thomas Tompion (1639 - 1713). The rivals were far from an amicable settlement. The king was ultimately asked to intervene. After weighing up the arguments, King James II awarded the patent to Daniel Quare, who won the dispute because his construction was more practical and more convenient. A single button triggered its striking mechanism, but two actions were needed to operate his competitors' invention. One button started the gong for the hours and another triggered the quarter-hour chime. The monarch was not pleased by the unnecessary complexity. Amongst other ways in which Daniel Quare proved worthy of the honor were to follow. In 1695 he presented a repeater that chimed 7½-minute intervals, i.e. 1/8th of an hour. The watchmaker Samuel Watson presented a pocket watch with five-minute repetition in 1710. The first watches that could ring the time to the nearest minute appeared in Merry Old England around 1750. Afterwards watchmakers strove to achieve further technical refinements and to miniaturize the mechanisms. Daniel Quare was irked by the fact that human error, such as failing to press the button all the way in could cause a repeater to ring the wrong time. His "all-or-nothing" mechanism assures that the strike train will either ring the time accurately or not at all. Julien Le Roy (1686 -1759), watchmaker to the court of Ludwig XV, improved this mechanism around 1740. This refinement occurred at about the same time as the disappearance of the mellifluous but relatively voluminous bells from the cases of repeater timepieces. Hammers which struck directly against the inner wall of the case freed space for larger, more reliable and more accurate movements. Abraham Louis Breguet is credited with the first use of meticulously tuned gongs. Subsequent repeater pocket-watches could be much slimmer.

SMALLER, SLIMMER, COSTLIER

The repeater's demise seemed imminent in the late 19th century, when prosperous people could switch on electric lights for illumination. But the repeater enjoyed a renaissance on the wrist, rather than in the pocket, beginning in 1892. Striking mechanisms for wristwatches had to be much smaller and slimmer, so they were miniaturized and thinned down. Watchmakers in Vallée de Joux and Geneva competed to reduce volume and thickness. Minute-repeater movements shrank to less than 3 mm in height with diameters of a mere 20 mm. The apex was reached with a diameter of a minuscule 13.5 mm in 1925. Further reductions weren't feasible. The shrinkage went hand in hand with higher prices. A 12-ligne movement (i.e. a diameter of circa 27 mm) for a wristwatch cost approximately twice as much as a 17-ligne caliber for a pocket watch.

The Swiss watch industry began offering accessible wristwatches with quarter-hour or five-minute repetition in the mid 1930s, but no brand achieved lasting success with its repeater. Nearly all such watches eventually disappeared due to insufficient demand.

This situation persisted until 1985 and the renaissance of mechanical watchmaking. Noble watch manufactures with the required developmental potential presented models that delighted fans of this complication, which is discreet because nothing but a little slide on the left side of the case reveals its usually mute presence.

Frédérique Constant and its Ateliers deMonaco have played in the mechanical upper class since 2010. Many manufacturers can assemble repeater watches encasing externally sourced repeater movements, but only an elite few manufactures can equip their watches with repeater calibers of their own provenance. A tourbillon added to this pretty picture is even more exquisite and is extremely rare.

Computer-guided fabrication of the ébauches is helpful, but not sufficient. Three hundred components, or significantly more parts than in the "Grand Tourbillon Minute Repeater" from the Ateliers deMonaco, must be finely processed and then laboriously and manually assembled to create a ticking microcosm. All this takes plenty of precious time. And as Benjamin Franklin, an inventor who is arguably on a par with A.-L. Breguet, aptly remarked: "Time is money."

DEMONACO
FEB
08
FRI
MONTH
WEEK
NEUTRAL
DAY
DATE
MADE IN MONACO

QP

CALENDAR STORIES

Were it not for astronomy and its quirks, the world of calendars would be a rather simple and straightforward place. It's common knowledge that our calendaric system is based on the Earth's orbit around the sun, but it's less well known that each yearly ellipse requires 365.24219879 days, which is equivalent to 365 days, five hours, 48 minutes and 46 seconds. This is the duration of the tropical year, i.e. the full cycle of the four seasons from one vernal equinox to the next.

Approximating this interval to 365¼ days prompted the Egyptian ruler Ptolemy III Euergetes to break a centuries-old tradition in 238 BC. His "Canopus Decree" transformed the time-honored calendar so that henceforth one additional day would be added to the 365-day Egyptian calendar every four years. Though astronomically correct, his decision was destined to be only a short-lived reform. Ptolemy III's successor rescinded the rule after only four leap-year cycles, thus returning Egypt's calendar to its accustomed form.

ATELIERS DEMONACO
dMc-QP-RR-WG Quantieme
Perpetual Calendar – QP-EZ
(2011)

Ateliers deMonaco

CHANGE OF SCENE: ANCIENT ROME

Each Roman year originally had ten months and a total of 304 days. The months of Januarius and Februarius were probably added around 700 BC. Twelve months of alternately 29 or 30 days each and one inexplicable extra day yielded 355 days, which equals one lunar year. Additional orientation according to the sun resulted in a peculiar calendaric hybrid which the historian Theodor Mommsen once scathingly described as "totally confused."

When Gaius Julius Caesar returned to Rome in 46 BC, what he encountered there with regard to calendars was anything but satisfactory. He eliminated the worst of the chaos by summarily inserting three months. The longest year in Western history lasted 445 days and was aptly termed "the year of confusion."

The dictator also initiated a fundamental reform designed to bring the rather arbitrary calendaric situation within his sphere of influence. The Greek scholar Sosigenes, whom Caesar appointed to serve as astronomical advisor, discovered in the course of his research that ancient astronomers had quite accurately calculated the year's duration as 365½ days in 4236 BC. Despite this fact, each year in the Ancient Egyptian calendar contained only 365 days. The annually accumulating error was periodically rectified by synchronization with the position of the star Sirius.

Arguably the ancient world's greatest astronomer, Hipparchus of Nicaea achieved an even more accurate calculation. He quantified the year's duration as 365 days, five hours and 55 minutes. Caesar and Sosigenes rounded this figure to the nearest quarter day and defined the year's length as 365¼ days.

The imperator's idea of circumventing the fraction by creating a cycle of three 365-day years followed by one 366-day year was analogous to the system stipulated by the Canopus Decree and may have been inspired by Caesar's Egyptian military campaign. Whatever its inspiration, the leap day became a determinative element of the Julian calendar.

ATELIERS DEMONACO
Caseback dMc-QP-RR-TI
Quantieme Perpetual Calendar – QP-EZ (2012)

08
DEMONACO
FRI
FEB
MONTH
NEUTRAL
DATE
MADE IN MONACO

DEALING WITH THE LEAP DAY

Caesar decreed that every fourth February would be lengthened by one leap day or "ante diem bis sextum Kalendas Martias." The start of each new year coincided with the customary assumption of office by Roman consuls, whose investitures took place on January 1.

Each odd-numbered month had 31 days and each even-numbered month lasted 30 days – except February, which had 30 days only in leap years and 29 days in ordinary years. What may well have been the Roman Emperor's most important political achievement took effect in 45 BC and at a rather unfavorable point in time with regard to the intended four-year cycle. Since leap years couldn't be determined on the basis of calculations, calendaric priests were giving the task of ensuring compliance with the predetermined cycle. The Roman senate recognized the consequences of this momentous calendaric deed and acted according to the maxim that honor should be conferred on those who deserve it. The month in which Caesar had been born was renamed to bear his clan name and thus Quintilis, originally the year's fifth month, became Julius.

Although the military might of the Roman Empire assured the rapid spread of the Julian calendar, its breakthrough can be credited to the Christian church because high-ranking clerics in Rome adopted fundamental parts of the Roman calendaric system. The choice of the year of Christ's birth as the zero point of the calendar is credited to Abbot Dionysius Exiguus, who concluded on the basis of the best available estimates that the 248th year after Diocletian was equivalent to the 532nd anni domini nostri Jesu Christi. That this isn't really true was first realized by the German astronomer Johannes Kepler, whose calculations showed that Jesus had actually been born seven years prior to the beginning of the new calendar. But a corresponding calendaric modification was inconceivable in 1606.

Let's return now to Caesar, the victim of a perfidious conspiracy hatched by Brutus and Gaius Cassius. The plot ended the controversial ruler's life, but fortunately didn't alter the principles of his calendar.

ATELIERS DEMONACO
dMc-QP-RR-RG Quantieme Perpetual Calendar – QP-EZ (2011)

THU
2
DEMONACO
DAY
MONTH
DATE
WEEK
NEUTRAL
44
48
52
4
8
12
16
20
24
28
32
36
40
MADE IN MONACO

A BIT OF REMEDIAL WORK

Misled by faith in the infinite perfection of his power, Caesar made his calculations without the aid of calendaric priests, who resumed the tasks of their office after his assassination. Unaware of astronomical backgrounds, the Pontifex Minor added an additional day every third year. Remedying this new calendaric confusion took twelve years under the aegis of Augustus Caesar. Whether intentionally or by coincidence, the new start of the Julian calendar in 8 AD gave mankind the opportunity to conveniently divide by four. The reason for a high honor from the senate, which also pleased Caesar's adopted son, was entirely another one: the conquest of Alexandria and the defeat of Antony and Cleopatra in Sextilis in the year 31 BC. In honor of Augustus' achievement, this month was renamed "Augustus" in 8 AD. Contrary to expectation, the monarch was not pleased. His month was the eighth in the year and thus an even-numbered one with only 30 days. Enviously viewing the longer month named after his stepfather, the regent demanded an improvement, which came in the form of a longer August and a loss of one day in February, the year's "leap month." Another consequence was the restructuring of the subsequent months from September through December. Thanks to Augustus, schoolchildren can now "read" the lengths of the months by counting them atop the knuckles of their paired hands.

ATELIERS DEMONACO
Close-up dMc-QP-RR-WG Quantieme Perpetual Calendar QP-EZ (2011)

REFORMING THE REFORM

If the commander who "came, saw and conquered" at the Battle of Zela on August 2, 47 BC and his advisor Sosigenes hadn't made a small error, the Julian calendar would probably still be in use today. But the negligible annual inaccuracy (a mere 0.0078 of a day) slowly accrued into a sizeable mistake. The timing of Easter in the Middle Ages led learned clerics to conclude that something must have been wrong. The Council of Nicaea had decreed in 325 AD that Easter would be celebrated each year on the first Sunday after the first full moon after the vernal equinox on March 21. But for originally unexplained reasons, Christendom's most important holiday slipped progressively later into the year. This couldn't be correct. The Holy Chair in Rome felt obliged to take remedial action. Research and reckoning discovered that the Julian year was 11 minutes and 14 seconds too long. The divergence between the stars and the calendar accrued to one full day every 128 years, so holidays assigned to a specific position in the calendar occurred later with each passing year.

Papal competence was needed to remedy this awkward situation. Popes Sixtus IV and Pius V died before the problem was solved, leaving Pope Gregory XIII to inherit it. A man of the Renaissance and keenly appreciative of the achievements of ancient Western and Eastern civilizations, Gregory XIII sought the long-overdue solution. He convened a committee of scholars, who presented their findings in September 1580. After taking ecclesio-political aspects into consideration, the pope signed the papal bull "Inter Gravissimas," which decreed that all Christians must purge ten consecutive days from 1582: Thursday, October 4, 1582 would immediately be followed by Friday, October 15, 1582.

Gregory XIII and his advisors naturally also devoted much thought to the calendaric future. The pope's reforms specified that three leap days would be struck from every 400-year interval. A secular year (i.e. one that commences a new century) would have a February 29th only if the year was also evenly divisible by 400: 1700, 1800, 1900, 2100, 2200, 2300, etc. ought to be leap years according to the four-year cycle, but they aren't because they're not by divisible 400 without leaving a remainder.

Incidentally, even the Gregorian calendar isn't absolutely accurate: it's 25.96768 seconds shorter than a correctly calculated solar year.

A CHALLENGE FOR WATCHMAKERS

Date displays first appeared on early geared clocks in the 14th century. These were soon followed by clocks with full calendars, i.e. indicators for the date, day of the week and the month. But these displays inevitably err sooner or later on timepieces with simple calendar movements. Due to the unequal lengths of the months, an ordinary date display remains correct for a maximum of 92 days, i.e. from July 1st to September 30th, whereupon a "31" appears on the dial to indicate the nonexistent September 31st, although October 1st began at one second past midnight of the 30th. The automatic display has fallen out of synchrony with paper calendars and accordingly needs manual correction.

Timepieces with perpetual calendars eliminate pesky manual resetting. Probably the first such pocket-watch is credited to the watchmaker Thomas Mudge of London. Completed in 1764, Mudge's masterpiece takes the lengths of the months in ordinary and leap years into account. Nearly all prominent watch manufacturers added perpetual calendar timepieces to their repertoires in the 19th century.

The first such wristwatch, a one-of-a-kind item, dates from 1927. Strictly speaking, it's a modified pocket-watch that was first made in 1898. Indicators that jump ahead at midnight are its special feature. Perpetual-calendar wristwatches were first manufactured in small series in 1941 and required manual winding. The world's first wristwatch with automatic winding and perpetual calendar followed in 1962. Speedily jumping displays rely on complex mechanisms and have remained veritable calendaric delicacies to the present day.

CALENDARIC WATCHES FROM THE ATELIERS DEMONACO

Chief watchmaker Pim Koeslag viewed this as a personal challenge. If the watch he designed was to be a perpetual calendar wristwatch, then it would surely be a very special one. As other complicated models with tourbillon and minute-repetition prove, normality has never been compatible with the lofty standards upheld by watchmakers at the Ateliers deMonaco. With this in mind, the "Quantième Perpetual" is an extremely rare wristwatch in which the automatic movement speedily advances all of the perpetual calendar's indicators at the stroke of midnight. The complexity of this achievement can best be understood by recalling that the ticking microcosm must survive on a mere one-billionth of a horsepower. The timepiece would grind to a halt if too much energy were suddenly sapped, e.g. to propel a calendar with large displays for the date, day of the week, month, leap-year cycle and – as a special rarity – also the calendar week. Conventional perpetual-calendar watches consequently require several hours to complete these daily changes. Their indicators creep slowly beneath tiny windows in the dial and are not completely legible for a lengthy interval before and after midnight.

QUICK SWITCHING

Frédérique Constant and its Ateliers deMonaco pursued an entirely different approach. Here, the calendar cadrature beneath the dial continually saps a very small quantum of energy from the movement and constantly transfers this power to a special spring which daily releases its stored energy at two minutes before midnight. Soon after the new day begins, all calendaric work has been done. The outsized date and the weekday display have jumped one position ahead. If Sunday has just turned to Monday, the hand on the calendar week display has also advanced. Events at the end of the month are especially exciting because the intelligent mechanism simultaneously moves several indicators. For example, the date display must advance from "28" to "01" at the end of February in an ordinary year. Pim and his colleagues have stored this program in a cam that annually completes one turn around its axis. The four-year cam turns 90° each year to signal the length of each successive February. If the owner of this watch wears it regularly keeps it in a watch winder when it's not being worn, he or she won't need to reset its calendar until the end of February 2100, when the cadrature, which obeys the traditional Julian calendar, will switch to "29." Of course, exceptional watchmakers of Pim's outstanding quality could also preprogram a mechanism to take into account the peculiarities of the Gregorian calendar, but the effort would far exceed its usefulness, which will become apparent in 2100.

SETTING AND RESETTING MADE EASY

The perfection of this patented perpetual calendar doesn't end with jumping indicators. Manual adjustment or readjustment of this function is rarely required, but this watch nonetheless has an uncommonly convenient and patented "EaZy" setting system. Adieu to the little stylus. Adieu to the attempt to bring its tip into position atop tiny buttons in the flank of the case that act on the cadrature. And adieu to the anxiety of fearing that a moment's inattention could put an ugly scratch in the precious case. Thanks to the "Quantième Perpetual" from the Ateliers deMonaco, all this antediluvian fumbling is now a thing of the past. A button in the crown preselects whether the next adjustment will affect the time, date, day of the week, week of the month or leap-year cycle. The user then turns the crown swiftly and without the slightest problem to reset the desired indicator. It would be hard to imagine a simpler or safer solution. Unlike many other models with perpetual calendar, this cunningly designed corrector mechanism is operable without problems at any time of the day or night. And unlike ordinary perpetual calendars, this one has no inoperative hours when it defends itself by blocking manual adjustment. The user also needn't worry about turning the crown in the wrong direction. This excellently legible wristwatch is uncommonly tolerant.

PRECISION AND RARITY

This manufacture-made automatic caliber and its hand engraved solid gold rotor are paradigmatic examples of contemporary precision watchmaking. Adorned with the coat-of-arms of the Mediterranean principality, the skeletonized oscillating weight winds the mainspring in both directions of rotation. The escape wheel is made of superlatively smooth silicium, needs no oil and collaborates with the oscillating system, which has a frequency of four hertz, to achieve rates that uphold the accuracy specified by the official chronometric standards.

Technical and horological refinement deserves a first-class housing. The round 43-mm case consists of 31 parts and in spite of its extraordinarily complex inner life, it is only 13.1 mm high. Personal taste determines whether to order the EaZy Quantième Perpetual in an elegant combination of titanium and 18 ct rose gold or titanium and 18 ct white gold.

Excellence is always a rarity, so the watchmakers in the Ateliers deMonaco will make only 88 watches in each of the two versions.

OSCILLATING MASS
Hand engraving of the Ateliers deMonaco 18K rotor

DEMONACO
MADE IN MONACO

FROM THE ATELIERS DEMONACO: THE FINEST TOURBILLONS

PULSING COMPARISON

Frédérique Constant has mastered the art of beautifully staging the heartbeat of mechanical wristwatches. This is also expressed via apertures in the dial through which connoisseurs can view the balance and the escapement. The tourbillon is a well-conceived alternative to conventional mechanisms that uniformly subdivide the flow of time. The dynamism of this little "whirlwind" is perennially fascinating.

ATELIERS DEMONACO
dMc-TB-RR-GR Tourbillon Ronde d'Or (2009)

DEMONACO
MADE IN MONACO

This was true for tourbillons in pocket watches, but the whirlwind began an unanticipated and nearly incredible evolution when it found its way into wristwatches. The tourbillon evolved in wristwatches despite the fact that this complication can only partially achieve its precision-enhancing effect on a wrist. To comprehend the reason, we must return to Paris in the late 18th and early 19th centuries, where we encounter a gifted watchmaker named Abraham-Louis Breguet, who is credited with several trailblazing watch-related inventions. One such brainchild is shock absorption, termed "parachute" by its inventor. A balance-spring with an upwardly curved terminal end is another of the master's brilliant innovations. In a nutshell, one could justifiably assert that the ingenious Abraham-Louis Breguet achieved more in just fifty years than had been accomplished by watchmakers during the preceding two centuries.

AN APPLICATION FOR PATENT PROTECTION...

The process of creating the tourbillon led Breguet from the workbench to the writing desk. On the 24th day of Floréal in the ninth year after the outbreak of the French Revolution, Breguet wrote: "Citizen Minister, I have the honor of submitting to you a note with the details of an invention which can be used in timepieces and which I have named the 'Régulateur à Tourbillon.' I apply for a constructive patent on this invention for a period of ten years. Via compensation, this invention prevents the errors which arise through changes in the orientation of the movement and through shifts in the position of the center of gravity of its balance.... Furthermore, I also rectify other errors which more or less significantly detract from the accuracy of the movement, and I achieve these rectifications in a manner which far exceeds the current status of our science.

After thorough reflection on these advantages, with the ability to perfect the production methods, and in view of my considerable expenses, I have decided to apply for a patent to certify the date of my invention and to compensate me for my expenditures."

The man who wrote these lines on April 14, 1801 had always categorically rejected compromises, so we can be quite sure that the tourbillon (the French word means "whirlwind") to which this patent application refers had already been realized with a high degree of perfection and functional reliability.

ATELIERS DEMONACO
dMc-TB-AC-GN Carré c'Or
Grand Tourbillon (2009)

ATELIERS DEMONACO
dMc-TB-RR-GN Tourbillon Ronde d'Or (2009)

ATELIERS DEMONACO
Ronde d'Or Grand Prix de Monaco
1965 – Unique piece (2012)

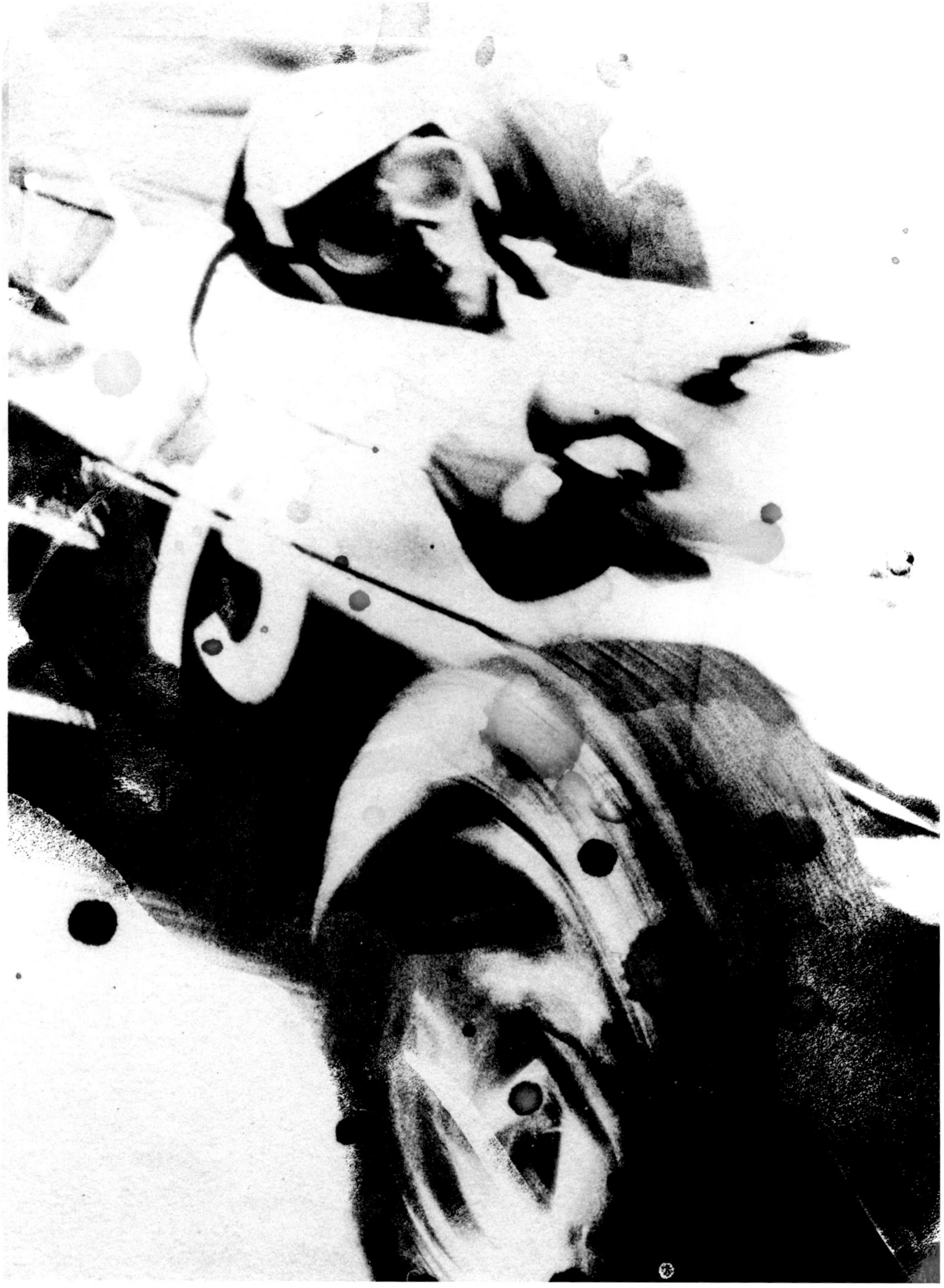

ITS TECHNICAL BACKGROUND

The reason for this technical innovation is provided by the Earth's gravity, which is an arch enemy of the precision of mechanical watches. The slightest eccentricity of the center of gravity of the balance and its hairspring, along with the varying frictional conditions accompanying these eccentricities, accelerate or decelerate the balance's oscillations. Watchmakers can counteract these irregularities by poising the balance with the greatest possible accuracy, but the familiar errors return sooner or later.

Breguet, whose clientele included high ranking nobles families and clergymen as well as wealthy bourgeoisie, was well aware of this situation. Unable to eliminate the problem, he devised a way to compensate for its inevitable recurrence. His compensatory solution was and still is astonishing for horological laymen and for Breguet's less talented colleagues. Put all rate-determining components (i.e. balance, hairspring and escapement) onto a toothed disc or inside a lightweight, optimally poised steel cage. Propelled by the movement, this cage turns once around its own axis during a predefined time span, usually one minute. The welcome result -- fractions of seconds that are lost, for example, during the first half of each minute are regained during the second half of the same minute.

This zero-sum game so deeply impressed the French minister of the interior that he granted the patent on June 26, 1801 (7th Messidor An IX) and recognized Breguet as the inventor of the tourbillon.

Then as now, mere mortals find it difficult to understand what this horological creation is all about. Most people in those days owned no timepieces of their own and were obliged to rely on clocks which their higher ranking contemporaries had affixed atop towers or on the imposing façades of noble edifices.

Then came the year 1806, or more precisely, September 24, 1806, when 1422 exhibitors began a four-week presentation of their latest inventions and achievements on the Esplanade in front of the Dôme des Invalides in Paris. Among the participants was A.-L. Breguet, who showed a "watch that compensates for all possible deviations in the rate of the balance and hairspring." True to its name, his creation stirred up a veritable whirlwind of admiration.

In 1923, British admiralty commander Rupert T. Gould wrote in his book about the history of marine chronometers that especially with this tourbillon, Breguet had provided the best proof of nearly superhuman skills and inventiveness.

HOROLOGICAL REALIZATION

Anyone who believes that a tourbillon reduces the technical and horological requirements needed to construct, fabricate and regulate a mechanical movement would be deluding themselves. Just the opposite is true. Only the best watchmakers can craft exclusive tourbillons. It takes consummate skill to make a complex "whirlwind" with the lowest possible weight. Tourbillons are unique, regardless of whether they dwell in pocket watches or inside wristwatches, where they first took up residence not quite three decades ago. But a tourbillon can only partially perform its compensatory function on the wrist, where its orientation relative to the Earth's center continually changes with the positions of its wearer's forearm. The best position for a portable watch is horizontal, where gravitation cannot disturb the poise of the balance. This is why the containers that house the movements of ultra-precise marine chronometers are cardanically suspended. No matter how wildly the ship may heel and roll, the chronometer's balance and hairspring always oscillate in an absolutely horizontal plane where the effects of gravity are negligible.

dMc Grand Tourbillon
2/18
ATELIERS DE MONACO
750
1/18
Titanium Gr.5

NATURALLY FROM ITS OWN MANUFACTURE

It was a sensational achievement when Frédérique Constant, a comparatively young brand, presented a tourbillon in the spring of 2008 that was developed and manufactured on the brand's premises. This family business had already been active as a fully fledged manufacture with several of its own movements prior to the aforementioned launch, but calibers with integrated tourbillons play in a different league than hand-wound or self-winding movements with conventional escapements. The tourbillon movement from the ateliers in Geneva's Plan-les-Ouates neighborhood was even more impressive. It embodied a series of unconventional constructive characteristics and the balance inside its tourbillon oscillated at the somewhat atypical frequency of four hertz, i.e. the rate-regulating organ completes 28,800 semi-oscillations per hour, thus permitting the gear-train to advance 691,200 times each day. In four or five years, after which this movement deserves a routine servicing, its lever escapement will have delivered more than a billion impulses of vital energy. The speedier frequency compared to ordinary balances results in a more precise rate and greater stability in varying orientations.

SMART WEIGHT BALANCING

Among the most outstanding special features of Smart Weight Balancing is its ability to assure that the load is optimally distributed inside the tourbillon's rotating cage. Eighty-one components make up this delicate assembly, into which watchmakers insert the pallet, escape-wheel, balance and hairspring. Each part is manufactured with minimal tolerance. Deviations from the ideal value must not exceed 2/1000th of a millimeter, but the technicians and their machinery generally achieve even greater precision. The lion's share of the parts are manufactured in a state-of-the-art, ultra-precise, computer-guided manufacturing center at Plan-les-Ouates in Geneva. These special machines do their difficult jobs with tolerances of 1/1000th mm in the X and Y axes and 2/1000th mm in the Z axis. Despite this extreme precision, it's nearly impossible to distribute the individual parts with absolute regularity inside the rotating cage. But when optimization of production is at stake, Pim Koeslag and his team are never without a solution that will ultimately lead to the desired result. Intensive thought culminated in the creation of the so-called "smart weight" system. Positioned on the outside of the cage, this innovation makes it much easier to poise the tourbillon, which, in turn, enhances the performance of the flyweight microcosm. Needless to say, the manufacturing is preceded by extremely precise calculations. The rotating cage is built with slightly more weight opposite the "smart weight" so meticulous fine adjustment can coax the center of gravity into the precise center of the tourbillon. The tricky feat of affixing tiny metal rings and miniscule screws, along with the repeated balancing out, consumes an entire day of concentrated work performed by a suitably experienced master watchmaker. (1)

(1)

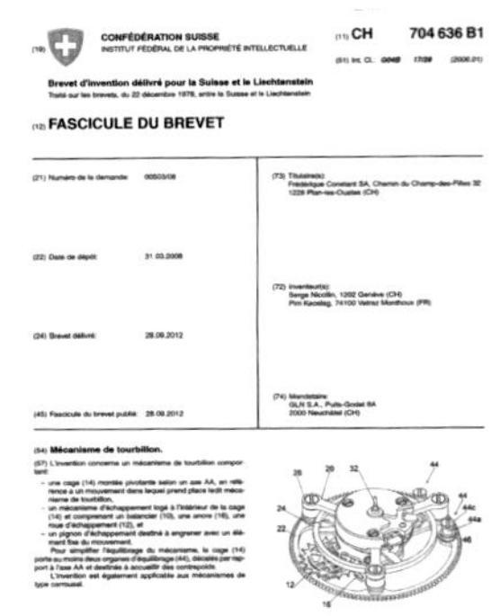
CONFÉDÉRATION SUISSE
CH 704 636 B1
FASCICULE DU BREVET

TOURBILLON
Drawing from the Patent of the Tourbillon Caliber

MORE THAN MERELY FINESSES

Adjusted and placed in the movement, the tourbillon begins to work neither too quickly nor too slowly, completing each pirouette in exactly one minute, as can be verified by comparing the seconds hand at its top with the seconds hand on a reference clock.

Perhaps the best way for its owner to appreciate caliber XP 1, which is made in the Ateliers deMonaco, is to observe the whirlwind through a loupe. Magnification reveals the fine processing lavished on all components, e.g. blued screws, polished steel parts, beveled edges and artful engravings. The two masters of their craft leave nothing to chance. They love their work, and they want their customers to see the ticking expressions of their affection.

No loupe is needed to appreciate the bearing bridge on the front side of the rotating cage. Such bridges are usually made from steel, which partially obstructs the view of the underlying tourbillon. This is not the situation in the timekeeping works of art from the Ateliers deMonaco. The undeniably high quality of this construction is embodied in a tourbillon bridge made from perfectly transparent sapphire crystal. This conceals none of the beauty beneath it.

All that now remains to be discussed is the indispensable escapement. The color of the escape wheel is impossible to overlook. Its bluish hue leads connoisseurs to suspect that it may be made of silicium, and indeed it is. The same material is used for the less visible pallet. Silicium, from which electronic microchips are made, became acceptable in haute horlogerie at the beginning of the 21st century. Monocrystalline silicium has the same crystal structure as a diamond. Silicium is 60% harder than steel, antimagnetic and resistant to corrosion. Even without post-processing, its extremely smooth surfaces make lubricants superfluous. Silicium is particularly advantageous because of its inherent lightness. Its specific gravity is 70% lower than that of steel. Especially in a tourbillon, great benefits can be reaped from every milligram of weight that can be eliminated from the weight of moved mass.

RONDE D'OR
dMc-MC-LG-RGD

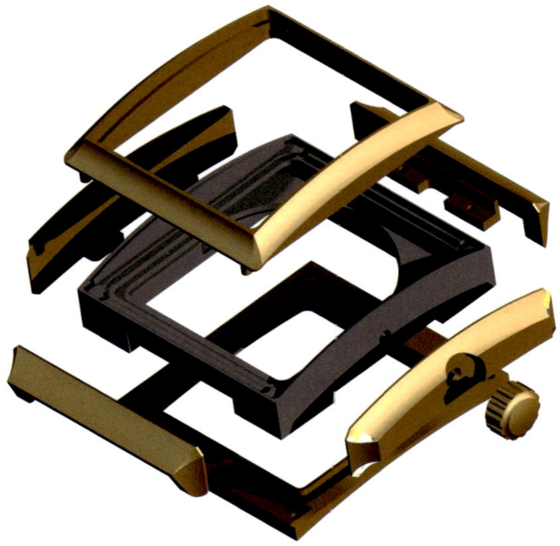

DEMONACO CASE
Technical construction of the Tourbillon Carre d'Or case.

HIDDEN BUT BEAUTIFUL

The back of a wristwatch is naturally hidden when the watch is on the wrist. That's a pity because the back of this watch is very attractive. However a practical folding clasp makes it convenient to doff the watch so that it can be inverted and its back brought into view. A sapphire crystal in the back offers a clear view of the skeletonized and decorated rotor. This oscillating weight winds the mainspring in both directions of rotation. The great strength of the steel spring keeps the movement in motion for 48 hours without any replenishment of energy. This ample power reserve is very reassuring. The components on the back of the movement uphold the same standards of processing and craftsmanship as their permanently visible counterparts on the front because Frédérique Constant and the Ateliers deMonaco are utterly uncompromising. Everything is absolutely first rate.

DECISION-MAKING PROCESSES

The agony of choice begins with cunningly crafted cases and elaborately designed dials. Alongside the round 43-mm-diameter "Tourbillon Ronde d'Or" with titanium core and white or rose gold mountings, horological individualists can also opt for an angular variant: the "Tourbillon Carré d'Or," which measures 44.3 by 48 millimeters. Here too, a lightweight titanium platform is adorned with valuable gold appliqués. The dials leave no wishes unfulfilled. Customers can choose from Roman or Arabic numerals, guilloche surfaces, or a combination of onyx and mother-of-pearl. It is hard to imagine any things more exquisite. Time consuming manual craftsmanship assures that every one of these ticking works of art is a unique artifact, so each tourbillon wristwatch is individually numbered.

FOUNDERS ATELIERS DEMONACO
From left to right, Robert van Pappellendam, Manuel da Silva Matos, Peter Stas, Pim Koeslag

LAST WORDS FOR NOW

We very much hope that you have enjoyed reading about the history of Frédérique Constant, Alpina and Ateliers deMonaco. Each of the three brands is still in full development. We are only now in the middle of our journey. Many new activities and beautiful watches are planned for the years to come.

We also hope that we have been able to convey the manner in which we live our passion through building the watch manufacture. We are committed to carrying on our long-term strategy of continuing to grow the company.

We keep these last words short. Our journey is ongoing. Hopefully, we will have more to say in the not-too-distant future. We invite you to follow our progress in the future. In the meantime, we strongly encourage you to live your own passion.

We would like to thank all those involved in the production of this book: Gisbert Brunner, Alexander Linz, Kevin Klatt, Gerald Guetat, Les Alchimistes, Marie Anthonioz, Yasmina Pedrini, Olivier Zaugg, Pim Koeslag, Hugo Herrero-Lévy, Benoît Favrat, José Santos, Manuel da Silva, Alexandra Sanchez and Pieter-Jan Stas. We would very much like to thank everybody at the company, our suppliers and most especially our customers. Success would not have been possible without your passion and hard work. We are blessed to have the opportunity to work with you. On behalf of everybody at Frédérique Constant, Alpina and Ateliers deMonaco, thank you for your interest and support. Live your passion!

Aletta and Peter Stas

INDEX

ACKNOWLEDGEMENTS:
Gisbert L. Brunner
Kindermannstr. 13
D-80637 München
Germany

Alexander Linz
Kreuzbrunn 15 / F5
A-3001 Mauerbach bei Wien
Austria

Gérald Guétat
Ars Aeterna Publishing
14, rue Charles V
F-75004 Paris
France

Chris Delker
6061 Village Bend, #3-0813
Dallas, TX 75206
United States of America

Kevin Klatt
Gorey
County Wexford
Ireland

LANGUAGES
Howard Fine English translation
Johnathan Franks English corrections

PHOTO CREDITS:
The position of the photograph or illustration is indicated by the page number, followed as necessary by the following abbreviations:
l = on the left, r = on the right, m = in the middle, t = a the top, b = at the bottom.

When there is no particular mention: Archives Frédérique Constant, Alpina, Ateliers deMonaco.

The photographs of watches, of watch details, events and of workshops, have been taken by:

Maurizio Bacci (New York – USA):
251

Jean-Paul Cattin:
91

Cessna Aircraft & PrivatAir:
233.

Karin Creuzet:
44-45.

Michael Emery – La Carrera Panamericana
119rt (x3), 120, 176, 178.

Dan Ferrer:
261, 262, 263, 264, 266, 267, 268.

Fleishman-Hillard Public Relations & Consulting (Hong Kong):
57t, 124b, 125, 132b, 135.

Hélice Classique Genève:
160rb.

Jess Hoffman:
28, 39, 42, 57m, 67, 70, 77, 78t, 88, 102, 103, 104, 106, 107, 108, 109, 110, 111, 112, 113, 119, 122rm, 123, 129, 132lt, 134, 136, 137, 140rt, 138, 140b, 150, 164, 168, 175, 179, 182tr, 183, 196, 206, 208-9, 216, 221, 230, 232, 244, 247lb, 256, 260, 261, 273b, 274, 297, 299, 302, 306, 314, 316, 318, 320, 324, 326, 328, 330, 335, 338, 339, 340, 343, 345.

Thierry Lagadec
272, 279b.

Christophe Lauffenburger:
26m, 27t, 38lb, 7tm, 84, 87, 90b, 215.

Lake Tahoe Concours d'Elegance:
123, 161t, 162.

Claudia Letule:
37.

Eric Maigret (Image & Co – Paris):
64br, 120, 121, 132rt, 181, 182tl, 182b.

Claire Matches:
253, 255, 257, 258-9.

Media Solutions Lab:
246, 247t, 248, 251-2.

Lies de Mol:
73 (Car and Boat background images).

Morand SA:
39t, 39ml, 39mm, 39mr.

Ludovic Di Orio:
Cover, 47, 52, 60, 61, 64, 65, 68, 69, 73lm-lb (Packshot products only), 77, 78b, 82, 83, 87, 92, 94, 99, 101, 140lt, 142, 144, 148, 152, 166-7, 223, 224, 226, 234, 236, 238-9, 241, 242b, 243, 248, 252, 276, 278, 284, 285, 287, 295, 303, 309, 310, 313.

Marc Ninghetto:
78mb.

Krzysztof Odziomek:
242t.

Robert van Pappellendam:
280, 291, 294, 296, 346.

Racing Team Level 5:
273t, 275.

Racing Team RLR:
279.

RenO – Agency les Alchimistes:
282-3, 292-3, 297t, 300, 301, 304.

Sebring Raceway:
272t, 272rb, 277.

Wladimiro Speranzoni (Venezzia – Italy):
165.

Henri Thibault:
156, 158-9.

Amy Tierney:
124t, 126, 128t, 130b.

Josh Wong:
6, 8, 10, 51.

Z-Visuel:
57t, 70mt.

AMBASSADORS:
ShuQi
Frédérique Constant Greater China
2009-2013
Rich and Famous Talent Management
4/F, East Commercial Block
18A South Horizons Drive
Ap Lei Chau, Hong Kong

Nicole Faria
Frédérique Constant World
2011-2013
92, 3rd Cross lavelle Road.
Bangalore - 560001
India

William Baldwin
Alpina World
2012-2013
Celebrity Consultants, LLC
3340 Ocean Park Blvd., Suite 1030
Santa Monica, CA 90405
United States of America

Aurélien Ducroz
Alpina World
2009-2012
Links Communication
1902, Avenue de Genève
74700 Sallanches

Romain Brandela
Alpina World
2008-2010
Romain Brandela
4 rue du Bois de Prunay
78910 Prunay le Temple

Level 5 Motorsports
Alpina World
2011-2013
5105 W. Clayton Road
Madison
53711 Wisconsin
United States of America

Nice for You
Alpina World
2011
Via Callalta, 1
31046 Oderzo TV
Italy

RLR Msport
Alpina World
2009-2011
Hall Moss Business Park
Bull Hill
Darwen
Lancs, BB3 2TT
United Kingdom

CORPORATE PARTNERSHIPS:
Cessna Aircraft
Alpina World
2011-2013
3 Cessna Boulevard
Wichita, KS 67215
United States of America

PrivatAir
Alpina World
2011-2013
Geneva Terminal C3
Chemin des Papillons 18
1215 Geneva
Switzerland

FREDERIQUE CONSTANT CONTRIBUTORS:
Marie Anthonioz, Yasmina Pedrini, Olivier Zaugg, Pim Koeslag, Hugo Herrero-Lévy, Benoît Favrat, José Santos, Manuel da Silva, Alexandra Sanchez Robert van Pappellendam and Pieter-Jan Stas.